TV

The Casual Art

TV

The Casual Art

MARTIN WILLIAMS

New York   Oxford
OXFORD UNIVERSITY PRESS
1982

Copyright © 1982 by Martin Williams

Library of Congress Cataloging in Publication Data

Williams, Martin T.
TV, the casual art.

1. Television programs—United States—Reviews.
I. Title
PN1992.3.U5W53   1982   791.45′75   81-3954
ISBN 0-19-502992-5   AACR2

Since this page cannot legibly accommodate
the copyright notices, the following page
constitutes an extension of the copyright page.

Printing (last digit): 9 8 7 6 5 4 3 2 1

Printed in the United States of America

Acknowledgments

This book includes both new and previously published material.

From early 1961 to early 1964, I regularly contributed a column on television to the *Village Voice*. My three contributions to the 1964-65 issue of *Kulchur* were also first conceived as *Voice* columns.

I was briefly TV columnist for *National Review* in late 1971 and early 1972, and contributed to *Listen* and the short-lived New York Sunday *Herald*. Meanwhile, I had contributed longer pieces to *Kulchur* and *Evergreen Review,* and later contributed on the subject to the *New York Times*.

This material is all collected here, with dates of original publication indicated. I have corrected some typographical errors in these pieces and re-inserted some dropped lines. I have also clarified and polished some points, without (I hope) unduly second-guessing my earlier opinions (except, that is, for a few foonotes I have added here).

Certain recurring ideas will be found in these pages and certain shows will be referred to often—"Gunsmoke" and "Dragnet" for two. I have, however, eliminated some repetition. My longer appreciations on Red Skelton and TV cartoons all but absorbed some earlier comments in the *Voice,* so I have not included those earlier columns. I have eliminated one *National Review* column for similar reasons. My longer comments on Dick Van Dyke were anticipated in "Thomas's Muffins" in the *Voice,* but in this case there seemed enough differences to warrant including both.

I was (and am) grateful to Jerry Tallmer, who was then the Features Editor for the *Village Voice,* for allowing me in. And to Barney Rossett and Fred Jordan of *Evergreen Review* for allowing me to write favorably on TV in so unlikely an outlet as *Evergreen.*

I am also grateful to the *Village Voice, Evergreen Review* and the Grove Press, *Kulchur, National Review,* and the *New York Times* for permission to include material that originally appeared in their pages.

Alexandria, Virginia   M.W.
September, 1981

Introduction

In late 1947, Budd Schulberg wrote that our film makers "have taken an instrument as sensitive, as delicately balanced, as capable of indescribable beauty and subtle emotion as the finest Steinway; they have set themselves before the largest audience in the history of the world—and have proceeded to play chopsticks . . . or a symphonic arrangement of chopsticks . . . in a spectacular production number that involves hundreds of pianos."

The words carried a certain authority. Schulberg had grown up in the movie industry. He was soon to become (with *The Disenchanted*) an established American novelist. And his statement appeared in the *Atlantic Monthly*.

His statement also carried an attitude that most of us were tempted to subscribe to. To be personal about it, I quoted his words in a column of movie reviews that I was contributing, as a twenty-two-year-old World War II returnee, to my college newspaper.

However, the mere availability of the modern Steinway does not produce the equivalent of Beethoven's opus 111, or Charles Ives's "Concord Sonata," or Art Tatum's *Aunt Hager's Blues*.

And the other problem with Schulberg's statement is that, within roughly a year of its appearance, Hollywood had offered or was to offer *Brute Force, The Yearling, The Killers, The Big Sleep, Boomerang, Body and Soul, Kiss of Death, Notorious, The Treasure of the Sierra Madre, The Search, The Postman Always Rings Twice,* and *The Pirate;* such high-grade slick entertainment as *The Bachelor and the Bobby Soxer, The Dark Corner, Nightmare Alley,* and *The Miracle on 34th Street;* and such examples of intelligent rubbish as *Crossfire, Possessed,* and *The Long Night*—and my list is only a random one.

Since the 1950s we have many times encountered the equivalent of Schulberg's statement directed at television. I hope that the words that follow will suggest two things to the reader. The first is that good television is not an act of the will, nor of good intentions, nor the result of the pressures of adverse criticism. And second, American television, finding its own way (like the movies before it), its own expressive resources, and its own means, has produced a share of drama and comedy that needs no apology, and that might be a source of satisfaction and even a basis for celebration, respect, and inspiration.

Contents

I

Theoretically

Tell Me a Story

Television, by a layman's rule of thumb, is a system by which pictures as well as sound may be broadcast. Television was at least theoretically possible in the minds of the early developers of radio. It was a fact by the early twenties. And it has been commonplace since the late forties.

Its early enthusiasts announced that it was possible by means of TV to bring about mass enlightenment. As an educational device, television might bring one expert teacher to hundreds of classrooms and thousands of students simultaneously. Used in the home, it might bring literacy to the adult illiterate. It could bring some history and a great deal of contemporary reality into the houses of the more sophisticated. In short, it could be a boon to education, formal and informal.

Such pronouncements about what man might do with TV were fairly predictable, and probably inevitable. The same sort of thing was no doubt said about the printing press. Within living memory, the same sort of thing was said about motion pictures and radio.

Perhaps it does not really matter much how anyone thinks television ought to be used. What does matter is how television is being used. And man has used television primarily in the same way he used the printing press, the movies, and radio—to tell himself stories, stories which purport to reflect his own condition and aspirations.

From one point of view, telling himself stories is man's most profoundly educational activity. But from the point of view of at least one critic, the highest fulfillment of television's potential came about when the prestige series "Wide Wide World" presented a simultaneous, live, split-screen shot of both the Atlantic and Pacific oceans. A dramatist might legitimately respond that such use of TV is neither educational nor enlightening but profoundly trivial. (How about a split screen of the Maine coast versus southern California waters for next week, Seattle versus Miami, the next—a whole series of split-screen shots.)

There are some who contend that TV should be used pedagogically, and they offer an approach which seems also to take a stance in favor of culture. It holds that TV is an ideal medium for the presentation of traditional and contemporary drama, which of course means drama originally created for the stage. Currently nothing is more critically prestigious for TV than to present a "special" which will pass for "serious drama." But any critical viewer of Sophocles or Shakespeare, or even Tennessee Williams, on TV must surely become aware that it is nearly as difficult to translate effective stage drama into effective television as it is to translate it into effective film drama. It won't do just to set up TV cameras in front of a classic play or an acclaimed Broadway hit. In addition, it may put gross limitations on television's own potential as a creative dramatic medium.

Historically, the first drama presented on television was written especially for the medium; it was a play called *The Queen's Messenger* broadcast experimentally on September 11, 1928. Its method of presentation was prophetic: three immobile cameras were used, each assigned to an actor photographed in close-up, and the resultant broadcast was a cross-cut of these three faces as they exchanged dialogue. In short, the technique amounted to a kind of photographed radio. And with important modifications, that is what television technique has remained ever since.

Understandably, Hollywood cameramen filming for TV hate the style—"just a bunch of tight close-ups and two-shots of people talking to each other," they are apt to call it. TV drama is not "just" that, but it very nearly is. Why, one wonders, does such a style persist despite the best intentions of actors, writers, directors, and cameramen who have worked in theatrical films, and who would like to present a more specifically cinematic style on television? The conclusion seems inescapable that, at its most effective, television is not and cannot be made a primarily visual medium.

According to one point of view, TV drama found its best expression in the early postwar days of the "live" New York dramas on "Studio One," "Kraft Theater," "Schlitz Playhouse," "Philco Playhouse," and the rest. Indeed, this period ending in the mid-fifties has been referred to as a "golden age" of achievement for American TV, an era before mass-produced dramatic mediocrity set in.

Partisans of this viewpoint might find it interesting to re-attend kinescopes of these early TV plays. In any case, it seems to me it is a critical mistake to attribute profundity to a quasi-naturalistic playlet like *Marty* in which a young Bronx butcher breaks away from the pressure of his family and friends long enough to marry a mother-surrogate schoolteacher. Or to attribute a devastating criticism of big business and its wheeler-dealers and philistines to *Patterns*.

Many of these early New York TV plays were formula dramas, and perhaps the formula itself, or at least the uses to which it was put, provides a critique of the ultimate merit of these shows, and a comment on their demise. In play after play, we are confronted with character A who has a problem, usually an unconscious problem. Perhaps the author thought of the problem as a flaw—dramatic if not tragic—but sometimes it seems little more than a bad habit. A is a spendthrift, a loud-mouth, a nag—whatever. A's friends and family are increasingly annoyed at his conduct and a crisis develops. So character B, a confidant of A, goes to him and says something like, "Say, A, you know, the trouble with you is that you always . . ."

A is quickly convinced that B is right. He seldom has any struggle with himself in admitting his shortcomings, and apparently he has little difficulty in dealing with them. His soul-searching is confined to a brief and facile conversation, sometimes argumentative and sometimes downright pleasant. ("Hey, I never thought of that, but you know you're right. I really ought to stop doing that.") And of course B's motives are never suspect.

I vividly remember a kind of unintentional parody of such plays which was broadcast toward the end of TV's "golden age." In it, A's flaw was that he was an unrelenting practical joker, thrusting exploding cigars, dribble glasses, and whoopee cushions at his male and female friends right and left, and loudly laughing his head off at the results. It got so bad that A even alienated his patient fiancée. The script treated A not as an offensive boor, nor as a tiresome but perhaps sympathetic bore, but as a terribly complicated young man with a significant problem.

Perhaps I am being unfair to an historically important period of TV drama. Or perhaps my criticisms are aimed not so much at those who wrote and produced the hour-long, New York-based dramatic shows as at the commentators who have championed the period as a golden age. Undeniably there were several good plays—Reginald Rose's *Twelve Angry Men* seems to me quite a good play. But the most significant contribution of live New York TV drama was, I think, no individual play, however good it may have been.

The major influence on the playwrights, directors, and actors who participated in this era of television was the Broadway stage, particularly the "serious" Broadway of the thirties, of Sidney Kingsley, Clifford Odets, Maxwell Anderson, William Saroyan, and the more contemporary Broadway manifestations of that tradition like Arthur Miller and William Inge. The outlooks and attitudes of these young TV playwrights, the style and the purport of their dramas, were by and large those of the New York stage of the time. The TV critics of the time were largely committed to those same attitudes and those same modes of drama, and the majority of set owners at the time were northeast and urban. The plays could not, I think, long interest and hold a mass audience of the kind that TV has today. And the somewhat haphazard and physically confining techniques used to present them (a scant handful of cramped studio sets) were too limited for the dramatic potential of the medium.

The contribution of this period lay precisely in the fact that it leaned upon the stage for its inspiration. Theater is of course a verbal medium. So is television, and these shows, if they did not discover that fact, at least underlined it. However, television is capable of a great deal more physical mobility and flexibility, and a great deal more dramatic terseness and condensation than a heavily stage-influenced style can offer. Television was to discover that it needed film, not to imitate the movies (but certainly to learn from them), and not (as with the movies) because film can be cheaply duplicated and hence widely distributed—one or two prints of a filmed television drama are quite enough to reach millions—but because film offered the controlled mobility and quick flexibility of dramatic image which would prove most effective for TV. Film can get around. And film can be edited.

One critic has remarked that perhaps a quarter of the shows presented on television are worth watching. And a successful TV producer said recently, with unusual candor, that he feels satisfied if ten shows a season of thirty in a given series are good ones, a figure only slightly higher than the critic's.

Both estimates seem quite high to me. What grand conclusions we might reach about the state of American literature if we decided that one quarter of the novels written in a given year were worth reading! Indeed, if one were forced to read only half the novels published in a year, he might easily conclude that the state of American letters should be described as a vast wasteland.

Probably most members of the TV audience—regardless of the height of their brows, the level of their intelligence, or the extent of their education—would concede that "most" television is poor. And why should anyone expect the situation to be otherwise? We take it as a matter of statistical inevitability, if not a matter of aesthetic fact, that most of the novels written in a given year will not be very good. And the same with most of the poems, most of the essays, most of the short stories, most of the plays, most of the music offered in a given year. But if most of the television offered in a year is poor, we feel something can be done about it if we hold an outraged congressional investigation, or at least bring it up at the National PTA convention, or perhaps sniff around suspiciously for some sort of conspiracy.

The question is not that most of us would consider most television boring—or at least that question does not seem a very fruitful one. It might be more fruitful to ask why such and such a show, regardless of critical opinion of its quality, has a hold on the public. Or, better still perhaps, how many really good shows are there on TV, and do they reflect techniques, qualities, and attitudes which one does not find in other dramatic media?

It has been said, of course, that television is not really a medium of drama but a medium of advertising. And more than one commentator has pointed to the disproportionate amount of time and money spent in the production of TV commercials as opposed to the production of TV shows themselves.

Assuming that time and money constitute some criteria for judgment (a dubious assumption, to say the least), is not the same sort of thing true of our magazines and newspapers? Juxtapose advertising with almost *any* activity, aesthetic or other, and advertising would probably prove to involve much more time and money. (It may also involve a great deal of expert craftsmanship, and the enormous influence of TV commercials on current cinema techniques—techniques which are usually considered avant-garde when they show up in a movie house—is a little-understood subject, but one which for my purposes here must also be another subject.)

Attacks on TV advertising take many forms. A man of considerable authority in American life once said that, on TV, sponsors sometimes influence the content of shows, but that advertisers in our magazines and newspapers would never think of doing such a thing. I submit that anyone who believes that is extremely naïve about what goes on in the editorial offices of American periodicals.

Some TV sponsors do take a direct hand in the productions of some shows. On occasion, the interference seems trivial—and trivial on both sides. A favorite story involves the writer who felt his art was being interfered with when he was forced by a cigarette sponsor to change the line "I'm lucky" to "I'm fortunate."* But when and if sponsors interfere in production, what they usually hope to gain is just as large an audience as possible. For in television we are dealing with a medium which supports itself, not so much by its advertising, as by its ability to please as large an audience as it can. TV commercials must bear the brunt of criticism simply by the nature of the medium: they are offered in time rather than in space. In reading a magazine, if we do not care what Virginform has to offer in its new line of padded bras, we can center our attention on the opposite page and peruse an article on the fallacies of academic testing. On TV we either attend what's in front of us or we turn off the set.

Well, then, what attitudes do please a mass TV audience? And are they the same attitudes that please the popular audience that still pays to go to the movies or that reads magazine fiction?

Let us take a plot: a young husband and his pregnant wife desperately need money. Fate tempts the husband with a situation in which it seems he can steal something and get away with it. However, he is discovered, and in an ensuing scuffle he believes he has killed a man. He flees, confesses to his wife and they become fugitives. They are pursued, overtaken, and the truth is revealed.

Such a plot might be treated in any number of ways. Ordinarily in recent drama—certainly in most 1930s' movies—the emphasis would be on the circumstances of the chase, the capture of the husband, and happy revelation that he is not a murderer. At a slightly

* A more recent variant on this attitude has to do with the effort of I.T.T. Corporation to raise its frayed international and national image by producing the prestigious children's show "The Big Blue Marble." While holding no brief for I.T.T., I get the feeling that, in the minds of some critics, a higher justice would somehow have been served if the Corporation had not produced the series.

more sophisticated level, there might be a note of irony: the husband is captured because he slowed down his flight in an effort to protect his pregnant wife.

On good contemporary television, such a plot would be given quite a different focus. We would not be asked to attach our interest to a complicated chase and capture, or to pin our hopes on a happy ending. On TV, the drama might easily focus on the psychology of the fugitives and their relationship: the effect on the young man's character of the fact that he has become a thief, and thinks he has become a murderer; the effect on the wife as she realizes that she has married a man who has become a thief, a murderer, a fugitive, and has made her a fugitive too. Anyone who has watched very much TV drama will know that it is possible for a TV show—a regularly scheduled TV "series" show, not a pretentious and prestigiously aimed "special"—to treat this basic plot for its dramatic essence as a conflict of character. What puzzles me is that so few of us seem to realize what an astonishing development in American culture such a treatment indicates.*

(*Evergreen Review*, June 1967)

---

* A more recent and more succinct example: the script to the Robert Donat film version of *The Count of Monte Cristo* applauds the hero's quest for revenge, finally in a glowing fade-out as the heroine walks beside him. In a recent made-for-TV version he and all his relationships are destroyed by his vengeance.

A Purple Dog, a Flying Squirrel, and the Art of Television

What we hear on television is more important than what we see on the small screen; in television at its most effective, and in contrast to movies, what we hear is primary, what we see secondary. But the visual images should underline, enforce, complement, integrate with what we hear.

The preceding is not my abstract hypothesis, but my effort to describe what happens on a wondrous television achievement: a group of thoroughly delightful, decidedly thriving shows, supposedly designed for children, and, at their best, unique comic experiences. They are "Huckleberry Hound," and its offshoots called "Quick Draw McGraw" and "Yogi Bear," and another (and better) series called "Rocky and His Friends." All of them are cartoon shows produced especially for television, and they are the only shows I know of that have a unique, well-realized television style.

One first comes upon these unpretentious little programs with a kind of awe and wonder, even before their fine comedy impresses him. The TV set seems bursting with creative life, an artistic energy and purpose one has not experienced from it before. It is almost like watching Griffith's *The Musketeers of Pig Alley* or a good Sennett chase sequence for the first time, and being almost overpowered with the unexpected expressive possibilities they have uncovered. These TV cartoons would seem so good nowhere else. Project them on a theater screen and they would pall; on television, their effectiveness is almost total. And they will soon make one aware of the wasted and mis-guided energy that goes into much TV production and, I fear, as-thetically dissatisfied with most of what he sees.

Economists would have a fine time rationalizing this effective-ness in their terms. TV cartoons have to be turned out fast and cheap, and conventional animation techniques take time and money. But I think the real reason is that any limitations can represent a kind of artistic challenge. Pantomime, camera, settings, and titles, these were what Griffith and Sennett had to work with, and prob-ably the real reason they achieved so much. If they had had sound, precise lighting, and other technical refinements they might have learned to express far less.

Cartoons (old movie cartoons) have been a staple on television since its beginnings. Ten years ago a few were being produced espe-cially for TV. Technically they were crude; aesthetically they were almost a succession of still pictures with a bit of movement, and the voices and scripts carried the show. Now, original cartooning for TV is commonplace; some of the results are good, some are terrible; some still don't move much. But, as I say, the best also discover the medium.

The drawings have to be simple and functional, the backgrounds fairly plain and stylized, the movements few and direct. Complex and difficult actions have to happen off-camera and be conveyed by crashes and yells. The little animals have to walk and gesture stiffly, and the artists are perceptive enough to make these stylized move-ments also amusing. About one thing the animators are very careful: the right lip movement for the speeches. The creative core of these shows lies in the writing, then with the highly resourceful people who do the voices—not to mention the sound effects. And I expect that all TV drama may be a writer's and actor's medium just as movie drama is ideally a director's. Accordingly, wild comic liberties are often taken with space and time—if a cat has a long, funny speech to make while running across a yard, the trip and the length of the yard are simply expanded to accommodate the speech. Always, the drawing keeps its place, and fulfills its place.

The "Huckleberry Hound" group is produced by William Hanna and Joseph Barbera, who used to do the MGM "Tom and Jerry" theater cartoons. Their style then owed a great deal to the wonderful cartoons made at Warner Brothers—Bugs Bunny, Daffy Duck, Tweety and Sylvester, etc.—surely some of the best American comedy ever done. The "Huckleberry" cartoons profit by such experience and such skill, but they have transmuted previous animation styles to the new limitations and purposes of TV.

All of the voices on "Huckleberry Hound," "Quick Draw McGraw," and "Yogi Bear" (and each show is made up of three segments with three sets of characters in each) are done by Daws Butler and Don Messick. They often use the time-honored cartoon practice of borrowing and adapting well-known manners of speech; Yogi Bear, for example, sounds quite a bit like Art Carney's Ed Norton. I should attempt some description of these shows and their characters but I think a random one will be best.

Quick Draw McGraw, for one, is a horse, a parody Western hero, and he has a Spanish-accented sidekick named Baba Looie. On his show there is also Snooper Cat, a trench-coated parody private-eye, who talks with Ed Gardner's "Duffy's Tavern" Archie malapropisms. There are Augie Doggie and Daddy Doggie. Daddy talks rather like Jimmy Durante and is constantly amused and ultimately chagrined by Augie's befriending and communicating with an assortment of bugs, ants, mosquitoes, other animals, and even Martians.

One sequence went more or less like this: Augie and Daddy are driving along. "Thank you, dear old Dad, for taking me on this picnic," pipes Augie. "Dat's OK, my son, my son," gravels dear old Dad. They pass a small duck with his thumb out who patiently explains to the camera in a Donald-ish quacking, "All the other ducks have flown south for the winter, but I can't fly because I'm too small and puny. So I'm trying to get there by hitch-hiking." Then, "Oh stalwart father of mine," says Augie, "there is a small duck beside the road seeking a lift."

"If there's one thing I can't stand," gruffs Daddy, "it's a hitch-hikin' duck."

Later, when he and Augie are trying to get extra rations from "generous, kindly old Dad," the duck announces, "Did you ever see a pitiful little duck stagger and faint from lack of food? Watch this!" as he ostentatiously staggers, spins, twirls, gasps, and finally falls.

Variously, we also meet Mr. Jinx, a cat, a sort of confused beat-nik Brando, who, like, chases Pixie and Dixie, two mice from down South. There is Snagglepuss, an enthusiastic lion who talks like Bert Lahr, reads his own stage directions ("Exit, guileless as ever, stage left!"), and acts out all the fantasies he can uncover: "I am merry Robin Hood, come a-robbing of the rich!" he announces to a police-man's annoyed "What are you, some kinda nut or somethin'?"

"Rocky and His Friends" is in some ways a more sophisticated affair, and it never risks the occasional calculated cuteness of the "Huckleberry" shows. It abounds in lampoon, burlesque, and wild puns. The basic stylization and simplicity in movements are there, but the drawing is less like movie cartooning and more like con-temporary comic illustration, and even an occasional BANG or HELP! is lettered onto the screen. The scripts are written in Hollywood, and for economy the animation is executed in Mexico. But there is a re-markable integration between the two, and Rocky also has some of the most rewarding visual delights on TV, particularly a stock pat-tern which closes each episode and shows Rocky the Flying Squirrel and his friend Bullwinkle the Moose suddenly and surrealistically sprouting up, blank-faced, in a daisy patch, under a burning sun.

Along with engagingly befuddled Bullwinkle, Rocky is, as the narrator explains in edgy tones that parody half a dozen TV news-men, "that jet age aerial ace, Rocket J. Squirrel," the cheerfully in-nocent hero of a serial lampoon. Their antagonists are Boris Badinov

("Hello there, fans of spying and sabotage!") and his bomb-throwing Baby, sloe-eyed Natasha Fatale. Boris, a master of disguises, has appeared as pilot Ace Rickenboris, Baby-Face Braunschweiger ("America's favorite gangster" and leader of the Light-fingered Five Minus Two), or Lord Chumley (with a heavily Slavic "Top hole, old bean. What ho!"). If he gets stuck, he can consult his *Pocket Book of Fiendish Plans,* speak with "Fearless Leader" by short wave, or, if he really goofs, get a visit from a shadowy "Mr. Big."

The "friends" of Rocky are "Fractured Fairy Tales" narrated with fine officiousness by Edward Everett Horton, or "Aesop and Son" told with eager innocence by Charlie Ruggles, and "Mr. Peabody and Sherman." Since I have an almost touchy respect for the wisdom of traditional lore, I hasten to say that the fairy tales are "fractured" on the Rocky show with a kind of loving respect, and lampooned without ridicule—even though every magic fish turns out to be a smelly, week-old carcass; every dwarf magician and tempter of the hero has the air of a fast-talking Bilko-esque press agent about him; and gnarled old witches are apt to sound remarkably like Marjorie Main and call everybody "dearie."

Mr. Peabody is, by his own admission, brilliant. He is a dog who has adopted a small boy, Sherman. Since he keeps him in a big city apartment, he has trouble amusing him and has invented a time machine, the Wa-bac, for that purpose. Through the Wa-bac, Mr. Peabody, with Sherman cheerfully tagging along, is able to help: a Chico Marx-accented Signor Borgia find an antidote against Lucrezia's arsenic-laden cacciatore; a confused and meandering Balboa to find the Pacific; a highly disinterested Fulton to getting his steamboat under way; and a slangy Sitting Bull to Little Big Horn in return for an autographed picture of Tonto.

Credit where it is due: "Rocky" is produced by Jay Ward and Bill Scott, Scott is the chief writer, and the voices are done mostly by Scott, William Conrad, June Foray, Paul Frees, Walter Tetley, and occasionally the ubiquitous Daws Butler.

The Fall will bring many new half-hour cartoon shows to television, but I do not particularly look forward to them. Ward and Scott will re-do essentially the same "Rocky" program in color as "The Bullwinkle Show," but it is hard to believe that Hanna and Barbera have already spread it so thin as to produce that humorless, derivative, noisy "Flintstones" show. The point, after all, is not necessarily cartoon shows, but the marvelous aesthetic language and style certain cartoon shows have been able to arrive at. I expect that what is needed now is for someone to realize the essence and the potential of this language if put to different dramatic purposes. Meanwhile, we are lucky—we in New York anyway—for "Huckleberry," "Quick Draw," and "Yogi" are each on once a week, and, counting re-runs, "Rocky" is on five times. That is about four hours of superior comic entertainment a week, and a lot to get from any medium.*

(*Evergreen Review,* Sept.-Oct. 1961)

---

* I should append a word about "Roger Ramjet," a 1965 syndicated cartoon series which spoofed such traditional radio-TV kiddy adventure shows as "Jack Armstrong" and "Sky King." Roger—square, rather pompous, and none-too-bright—led Yank, Doodle, Dan, and Dee ("Ramjet and that bunch of rotten kids," a villain once called them, in the voice of a deranged Burt Lancaster) in a never-ending battle against the sinister forces of evil. The animation was bare-minimal, and some outrageous anti-cinematic jokes were played with it: talking profiles were "flopped" (reversed) in the middle of speeches; the same shots were amiably cycled and re-cycled, sometimes in the same episode. Again, the scripts and voices carried it all. And the writers seemed to be having a wonderful time. One segment made constant (but sanitized) allusions to the punch-lines of several standard dirty jokes. Another episode had a time machine plunk Roger, in his jet togs, down in Renaissance Florence. Seeking directions, he knocked on the door of "Leonardo, Painter," to be greeted by an ebullient, "Hey, kid, a-come on in! You wearing a funny suit, you look like a swinger! Go home, sweetheart!" he adds, to his female subject (who of course looks quite like the Mona Lisa), "I paint-a your pitcha tomorrow." Well, in any case, a cartoon series with very little animation turned out to be most enjoyable—and that is a point for proponents of full animation to ponder.

Changing the Subject

"Laugh In" overtook us in mid-season 1968 after a couple of success-ful pilot shows, assaulting us with a barrage of one-liners, quick blackouts, graffiti painted on undulating female thighs, song frag-ments, and absurd pratfalls in which grown men in rain slickers fell off tricycles that were barely big enough to hold a five-year-old. It kept up such assaults for three more seasons, and made most of us like it. And with too many cast changes, and a major change of pro-ducers, the show somehow hung on for two seasons more.

Who could follow it, this gleeful, breakneck barrage, whose zingers went by so fast that we never quite caught the last one be-fore we were in the middle of the next? Well-behaved families had to set up rules to restrain their intrigued but puzzled offspring: no questions asked during "Laugh In," and no jokes explained until the commercials or the end of the show. Take the time to explain one joke and you miss the next six or eight. But we liked it, we loved it, we were dazzled by it, whether we understood half of it or not.

How could that be? I think because, besides the jokes and sight gags we did understand, we sensed that the show celebrated and parodied something fundamental about the nature of TV itself, and we loved that discovery and celebrated the parody.

Television demands a terseness and a condensation beyond that of any other medium of comedy or drama. Movies, with dialogue that underlines and complements the visuals, call for shorter scenes and a quicker momentum than the stage. TV, with visuals that underline its words, demands more condensation still.

TV, once we get used to it, can make other media seem padded sometimes. James Bond decides that he must confront Dr. No in person at his island headquarters. We watch Bond approach the waterside in long shots, see him haggle for the use of a boat in medium shots, watch him *put-put* across the water in long shots, see him disembark and approach the doctor's mansion in more long shots, see him stopped at the door by a housekeeper-guard and talk his way in with medium shots and close-ups, to be ushered finally into Dr. No's study. The more affable Man from U.N.C.L.E. would have made his decision a signal for a direct cut through space and time to Joseph Wiseman's malevolent Dr. No face, pivoting into the camera from behind his desk.

It may be futile to ask why TV calls for such condensation to be its effective best. Whatever the reason, it simply does. Actors must often wish that it did not, for it calls on them to make some very difficult emotional transitions in some very terse and short scenes. An actor may have to go from outrage to trust, an actress from fear to love, in only a few lines of dialogue, and be convincing about it.

I think that one reason that music has never been successfully presented on TV, even the best music with the most sympathetic cameras observing it, is TV's short attention span. And the quality of the music aside, perhaps the presentation of pop singers and groups on TV variety shows offer the best answer: give the number its own setting, keep its production and its contact with the cameras direct, keep it relatively short, show it to the audience at home with no sham, no apologies, no cut-aways to dancers unless it's a dance you're doing and not a musical number, and then go on to the next sketch, number, whatever.

It should probably not surprise us that the most effective and justly praised programs on "public" TV, "Sesame Street" and the "MacNeil-Lehrer Report," are those that have learned, and learned well, the implicit lessons that commercial TV, with its far more experimental nature, has to teach. And the lessons are: keep it brief, keep it direct, keep changing the subject (or *seeming* to change it—something which "MacNeil-Lehrer" does very well), and keep changing the scenes or the setting.

II

Chronicles

Seasons by the Set

Soup for Lunch

I have a lot of trouble with the TV shows I am supposed to like, but I have one pretty good guide. Whenever I hear that the PTA has put anything down, I always look into it. A few years back I discovered "Gunsmoke" and "Have Gun, Will Travel" when the PTA attacked Westerns, and those shows have a stubborn moral energy beyond anything you have a right to expect. On their latest pronouncement, I beat the PTA to the punch. They went after a man named Soupy Sales, muttering something about his misuse of the English language, but I have been watching Soupy with great delight for a long time.

"Lunch with Soupy Sales" is a highly informal show intended for children. On the face of it, Soupy is a personable young collection of wonderful old jokes and dance steps, dressed in khaki pants and black sweater, and topped by an innocent and generously mobile face that is pasted at least twice in each show with a gooey pie. The pies are usually thrown by White Fang, "the meanest old dog in the United States," a huge hound (yes, a pie-throwing dog) never seen on camera except for the very expressive length of his right forepaw. Soupy has another dog named Black Tooth, "the sweetest dog in the world . . . don't kiss," who is equally big and equally unseen except for a left forepaw.

These energetic paws, plus an assortment of gesturing right and left arms supposedly belonging to a succession of irate neighbors, crafty door-to-door salesmen, and adamant delivery clerks, are actually parts of the ingratiatingly talented Clyde Adler. With his arms and hands and a few high and low growls and shrieks that are vaguely analogous to English words, Adler manages to give both of the dogs such rompingly lively personalities that their recent, wild, off-camera foray through a supermarket was entirely vivid to the audience, although it was conveyed solely by Soupy's disbelieving reactions to a store-manager's panicked telephone call.

Adler also manipulates Pookey, a *very* small lion, and Hippy, a baby hippo, both of them hand-puppets with the most engagingly humorous personalities I know of on TV this side of Carol Burnett. Watching Pookey communicate with Soupy through Adler's shrill whistle and bobbing right hand can be a delight. Seeing his gleeful, burlesquing pantomine to a rock-and-roll record, as Soupy offers Greek-chorus-like encouragements, is a wonder. Those occasions when Soupy reads a chapter from the Grimm Brothers, as, to his rear, Pookey secretly and mockingly acts out the parts, complete with rapid changes of costume, is a surpassing joy. Soupy, Adler, some wild sound effects and scattered film clips—these are the show.

There is one strange moment each week when Soupy delivers some "words of wisdom," advice on being good to mommy, written in a strained doggerel. But Soupy rattles off these almost unintelligible verses with the same self-effacing charm that makes the quaint jokes he uses so funny. His is a genuinely warm and innately amused personality, and he has one of the few live TV shows I know of whose basic ingredient is not its proprietor's inflated ego. Anyway, if he ever steps out of line you can be sure that White Fang will let him have it right in the face with a well-timed pie—to nobody's surprise but Soupy's.

(*Village Voice,* March 30, 1961)

Waiting for TV

Last week's "Play of the Week" announced "A Cool Wind Over Living" as a "brilliant new drama" about today's troubled generation, but its characters seemed more troubled by some Noel Coward-style dialogue ("What if there really is a god?"/"Don't be vulgar" or "Psychiatrists invariably have bad taste and live on the Upper West Side"). This week, in "Waiting for Godot" there was the promise of a theatrically fascinating, even stirring, play, and casting that looked provocative. Yet I gradually found myself with the same bored lethargy as last week.

This week we can't blame the play. We might blame director Alan Schneider's quite inappropriately hurried staging; or his odd, ineffectual blocking—front-to-back, away from the camera. Of course on the surface the play is a kind of vaudeville, and there is a cast that seems right for it. Zero Mostel's skilled comic theatricality has become stock-in-trade for such productions, but somehow he hardly gets going. Kurt Kasznar's usual flair seems closer to an indulgent camp in his Pozzo. Burgess Meredith behaves with unaccustomed modesty for the present stage of his career, but this does not make his Didi seem potentially a Pozzo, as E. G. Marshall's did seem on Broadway. And possibly because he was asked to move around a lot, Alvin Epstein's formerly pathetic Lucky seems forced and shrill.

But in the theater such flaws might not induce passivity or boredom. The answer may be as simple as this: as the movies discovered long ago, you can't do something aesthetically virtuous just by setting up cameras in front of even the best stage play.

Notes: No sooner had I written my first column for the *Voice* and praised Soupy Sales than his program was canceled by ABC. If noon Saturday still seems a good time for comedy, and you don't require yours to be intentional, try "True Story." The last one had more lurching, arm-flailing, eye-popping, and set-chewing than an old Marie Dressler comedy. It is "live TV drama from New York," and anyone who finds magic in that phrase should see it and ponder.

(*Village Voice,* April 6, 1961)

Information, Medium

The "Circle Theater" is one of the last vestiges of "live TV drama from New York" and something of a showplace for television as a medium of public information. It began life as little brother to the hour dramas, and it was pretty soapy in those days. I remember one about a girl who decided to give up that piano-playing sharpie and return to her minister father's bosom. (Yes, Electra, you *can* marry your father. This is America!)

Nowadays the "Circle Theater" offers quasi-documentaries on current events, problems, or just plain curiosa. Recently there have been looks at black-market babies, term-paper ghosting, gifted children, the new respectable look in gangsters, and this week they offered a revised version of the one on how Eichmann was tracked down. The shows are usually highly entertaining and always remarkably skillful.

True, the acting is frequently perfunctory, but there is probably little time for any but technical rehearsals. The real skill comes in the sheer physical coordination of the show, and a staggering job is done in the control booth. On a recent look at the activities of Interpol, the show traveled over half the globe on land and sea. Literally hundreds of shots and camera set-ups, film clips, scores of sets, and dozens of actors were all rapidly manipulated, juggled, and swiftly integrated into a smooth, absorbing narrative—inevitably (this is TV) an intrigue story of dope smuggling. Very much a part of the success of these shows is narrator Douglas Edwards, who can make almost anything seem at once important and mellifluously comforting.

I have no idea of the accuracy of these "Circle Theater" reports. I know that a recent description of heroin addicts implied that they are depraved and violent, and that doesn't fit with my experience. And that on a show about fortune-tellers what was supposed to be policewoman's headquarters looked more like a YWCA office. But accuracy or fairness has nothing to do with the pleasures of watching these little melodramas. The program uses a kind of *Reader's Digest* approach: a problem is posed and made to seem terribly important, pressing, and even dangerous. As the tension is built, gradual reassurances are introduced, and we end with a fuzzy glow— someone is looking out for it, everything is going to be all right. Every problem becomes a diverting little suspense thriller with a happy ending, and the viewer can reassure himself that he is a responsibly informed citizen for having looked in.

The show is produced by David Susskind's Talent Associates, and I guess you know he comes on pretty strong about TV's responsibilities.

(Village Voice, April 13, 1961)

Salvation by Statistics

Last Sunday the CBS documentary "Junkyard by the Sea" was re-run by public request, and again there were all the shocks and terrible statistics that a trip through the Tombs, women's prison, and Riverside Hospital can provide. There was Commissioner Anna Kross, saying that jails were no answer but institutions were. There was the docile, complacent addict (interviewed in impatient, leading-question TV style), mouthing psychologists' commonplaces about her "case" and "society." There also was the strikingly honest female addict wondering how the guards, nurses, and doctors could stand by during the enforced suffering of withdrawal day after day.

There was Commissioner Anslinger, his face revealing more of his real attitudes than his few words about addicts as Typhoid Marys who should be quarantined in hospitals. There were the patients at Riverside Hospital, the subjects of directors' meetings. There was their sculpture and painting with its undeniably passionate recurring symbolism. (Was there a Jungian in the house? I hope so.) There was the group-therapy session with doctor and patients apparently so self-indulgent. There was the presiding authority who said that if a patient goes out and works for a while or limits himself to one shot a day, "we think we have success," adding that there is no magic cure—yet.

Then the institutionalized hopelessness suddenly stopped. With the East Harlem Protestant Committee there was some kind of help with no paternalism and no need or desire to judge attached. Norman Eddy revealed a moral commitment in his few minutes. "We began"—these are more or less his words—"because we saw so much suffering around us. And because we believe in the individual human being and believe in trying to alleviate suffering." And then he said that anyone who looks for panacea or magic solutions misunderstands completely the problems of addiction. He gave no statistics, no percentages, no threats of expenditure—and no government agency could ever say what he said: the problem is individual, and the only way to help comes from a genuine desire to help the individual.

In the end, we went back to the recommendations about research and public projects, so necessary—and so easy to talk about.

(*Village Voice*, May 4, 1961)

Children's Hours

We are constantly told that television is an ideal medium for public information and debate, and are now officially told so by the FCC. We have heard such things said before, of course, about the movies and about radio, and I suppose it is inevitable that we should want to turn media that are getting the attention of enormous numbers of people into media of education and enlightment. But I wonder if it is not precisely because they are such *mass* media, so very collective, that films, radio, and television so rarely do a job in keeping people informed and in getting issues discussed. The one really developed medium of information and debate we have is—for all its limitations—the public press, and for every TV network there are hundreds of newspapers and magazines.

But what matters, of course, are results and not my theories. In my early youth I heard lots of radio news and saw lots of newsreels and some much-praised film documentaries. I acquired lots of prejudices and a careless ragbag of facts, but I certainly never became really informed about anything.

On television I am as grateful as the next man for live coverage of an important U. N. debate, but what about programming that takes some production and planning? The few documentary and "discussion" shows that are currently on TV are highly praised and are apparently considered good models for the future shows. For example, there is "The Nation's Future," produced by Irving Gitlin. Mr. Gitlin has high-mindedly said that " 'Gunsmoke' . . . will do nothing to affect your future or that of your children."

"The Nation's Future," on the other hand, recently asked Otto Preminger and Dore Schary to discuss movie classification and censorship. Schary and Preminger might be interesting on the subject (interesting to Leonard Lyons anyway), but I don't suppose that former Postmaster General Summerfield would be any less qualified to clarify the issues involved.

Actually I hope Mr. Gitlin is wrong and that my children are deeply affected by "Gunsmoke," because, for all his swaggering, Marshal Dillon is about as close to a tragic being as any popular entertainment in this country has ever offered. Imagine: a Western hero who has moral doubts and decisions to make and who knows he can be wrong. And weekly vignettes which are often moving enough to leave one pondering just where the right and the wrong were.

"Gunsmoke" often sees humanity with a compassion I have never encountered in a "public-affairs" discussion and never expect to.

Of course there are those precious moments of low comedy on news shows. For instance, don't miss the almost nightly exchanges between Gabe Pressman and Mayor Wagner on the 11 o'clock news.

(*Village Voice*, May 13, 1961)

Ernie's Way

A few years back Ernie Kovacs did a television ballet, a version of "Swan Lake," quite legitimate except that his entire corps of dancers was dressed in gorilla suits. If such a thing had happened in Paris in the 20s and 30s, it would surely have been considered a devastating artistic coup. Here, it is one of several evidences one might cite that Kovacs has a very original and imaginative comic perception.

Last Thursday he presented a half-hour of vignettes, comic and serious, set to music of various sorts. It included a great deal of what the trade calls "visual comedy," but the phrase doesn't say much about a show which began with an adenoidal, off-camera voice singing "Mack the Knife" in German while we watched the frequency impulses it set up wiggling on an electronic tube. Every minute or so, as "Maki Messer" droned on, we cut briefly away to other sights: the water drained rapidly from a bathtub and a human hand sought desperately for a hold before disappearing through the pipe; a young man gaped open-mouthed as the soda went out of his glass, although his date was clearly taking drags through a straw stuck in her own; a man entered a photographic darkroom "live" only to leave it a moment later "on film" and in negative. There were also (by my count) six variations on Loretta Young's sweeping entrance, including one where she got a handful of loose doorknob, another where she got a face-full of loose pie from "John," and a third where she was utterly horrified to discover John, doing we never did see what, across the set.

There were also some gently funny bits of people eating in time to spirited martial airs and the conversation of two apes in a space capsule. Foils to these were a series of abstract, kaleidoscopic patterns set to Bartók, and a condensed view of a big-city back street from dusk to dusk, which, chiefly because it attempted only borderline realism in its sets and its people, was better than many an arty documentary of its kind.

Someday there may be fairly earned studies of Kovacs and his Superclod or Percy Dovetonsils. Meanwhile, he is such an expert comic actor that he can spend most of his time subtly gracing films in character parts. That, surely, is TV's loss. If Kovacs feels the loss too, we may see more of him.

(*Village Voice*, May 25, 1961)

Whose Wasteland?

It is probably time for some sort of statement from me about how I am going to occupy this space. In the first place, I am not going to waste your time and mine by frequently reminding you (one way and another) that most television is bad.

Second, I will not tell you much about what ought to be on television because I don't know. I expect that at this infant stage anything ought to be on TV that somehow explores the medium and finds out what will work effectively on television and what won't. In this respect Phil Silvers, and even Ed Sullivan, have made at least as much of a contribution as "Camera 3" or "The Open Mind."

At any rate, whatever anyone thinks ought to be on TV, what is on TV is mainly (1) drama, serious and comic, and (2) variety shows. I think it is best to remember that TV has replaced the movies as *the* popular medium, and remember that the movies were Richard Dix as well as John Huston, that Chester Morris was sometimes a better actor than Clark Gable, and that some of Wild Bill Elliott's cowboy pictures were worth anybody's time—that TV must replace the quickie crook and cowboy shows as well as the MGM sheen.

Those things being granted, I think TV does remarkably well. On Tuesday, for example, I can see the Sgt. Bilko re-runs, take a chance on Hitchcock's gimmick plots, take another chance on "Thriller" (where the scripts are apt to be poor but one frequently encounters superb acting by virtually unknown players), and see Garry Moore manage a remarkably consistent variety show.

Television simply must provide a lot of trash, and when the trash is as generally well done as "Perry Mason," "Checkmate," "Hong Kong," and "77 Sunset Strip," I see little point in flailing at "Hawaiian Eye" or "Malibu Run."

It is nice that FCC chairman Newton Minow apparently plans to watch TV, but calling it a "vast wasteland" is a little too easy. What matters is Minow's idea of a flowering cultural garden. After all, those tawdry "Steel Hour" plays produced by the Theater Guild are obviously somebody's idea of culture. But if more of that sort of television is one of the sacrifices the New Frontier asks of us, I'm copping out.

Note: Another top level pronouncement comes from Robert Wood of NBC, and is about the many 60-minute shows planned for next season: "Hour shows allow for [character] delineation and don't have to depend on violence." Well, let's see: "The Untouchables" is an hour show. So is "Surfside 6." "Have Gun" is a half-hour show. Got it?

(*Village Voice*, June 22, 1961)

Cracking Heads

Anybody who wants to see violence on TV can get plenty of kicks watching "Victory at Sea." At least that is what a reader says, with enough examples from recent re-runs of the series to be thoroughly convincing.

Of course there is violence and there is violence, and such things should be discussed only with the greatest aesthetic, cultural, and psychological care. But they are being discussed as if violence in art is always automatically deplorable, as if violence in popular drama were a cause of something and not a symptom, and as if life always follows art in such matters. (And also as if those "Studio One" and "Kraft Theater" days were a "golden age" of TV—but that is another story.)

The television screen is small and (as they have been telling us for a long time) its effect is largely intimate. It is hard to convey the excitement and scope of a chase or acrobatic escape or the other staples of movie "action" melodrama on TV. The temptation to substitute a fist in the face for a run across the rooftops is almost built into the medium.

Some of the violence in current television (and current movies and current fiction) is extreme, gratuitous, and aesthetically unpurged. It is piously denounced, but to what end? It seems, for example, that in the name of protecting children from excitement and violence we have attempted to raise nearly two generations on "Little Johnny Goes to the Supermarket" instead of Homer, the Grimm brothers, or almost any appeal to their imaginations. Meanwhile, a disturbing violence erupts from their crime and space comics, from TV screens—and in the streets. Cause and effect?

It is quite possible that the most deplorable thing on TV is the complacent, facile, smugly superior "morality" of the "Loretta Young Show"—but don't expect any high-minded investigations of that one.

(*Village Voice,* July 6, 1961)

TV Culture, at Random

Well, the congressional hearings on television held in New York are over. At first the word was that they could have no immediate effect on programming or scheduling, that next season was already set anyway. Now it seems that it is not too late and there will be changes.

Many people seemed to have been shocked by the revelation that boorish things go on in TV producers' offices. I can only assume they don't know what goes on in theater producers' offices, movie producers' offices—not to mention book, magazine, and newspaper editors' offices. To be sure, there were some shockers, and perhaps most shocking was the display of facile conceit and cultural snobbery on the part of most of the witnesses. The clincher came from Fred Coe, who used to produce those look-ma-I'm-being-deep hour dramas which seem to be every TV columnist's idea of great art. Coe said that TV is "a medium of extremes" with " 'CBS Reports' at one end and 'Maverick' at the other." (Which end is up, Mr. Coe?)

The producers asked for violence and sex, it seems. A writer could run away from such a request, of course. Or he could corruptly embrace it, à la Mickey Spillane. But what should a realistic moral drama in our time deal with? And how would a really responsible writer respond? And what do some of our best writers deal with? Violence and sex maybe?

I note that one columnist has expressed shock that a show called "Brenner" deals with a cop who investigates other cops. He feels that's somehow immoral. Another TV reviewer strayed from his field long enough to be shocked at Otto Preminger's statement that he might soon be making a movie about a man who has an affair with his mother. (It's been done; it's called *Oedipus Rex* and I saw it on TV.)

I have already had a nightmare vision on the 1961-62 TV season as a middlebrow wasteland, presided over by David Susskind, Jack

Gould, and John Crosby, with interview shows where Sloan Wilson, Vance Packard, and Al Capp make it as intellectuals, and "Patterns" and "Marty" are the models for TV plays of deep moral commitment. For some idea of what it might be like, limit yourself only to the "Playhouse 90" re-runs and the shows *Life* magazine recommends in its weekly list.

But, then, that makes me a cultural snob, doesn't it?

(Village Voice, July 27, 1961)

Odorous Comparisons

The horse may be dead but the breed isn't, so perhaps something can be gained by flogging it a little—a show called "Way Out" has been cancelled, but the fact that it appeared at all suggests a few things need airing. "Way Out" was done in New York "live" (i.e., on tape) by local talent and produced by Talent Associates (and most of what I have to say about it also applies to the same outfit's "Great Ghost Stories"). It is, like it or not, a kind of test case for New York's continuance as a TV production center. And if "Way Out" is a sample, New York has little to offer.

In the first place, the program was very derivative, a kind of cross between the Alfred Hitchcock show and Rod Serling's "Twilight Zone," complete with Roald Dahl giving sophomoric imitations of Hitchcock's urbane, showbiz introductions.

Of course you take chances on the Hitchcock and Serling shows. Especially on the former, the gimmick is the real subject, so the shows come off divertingly or they do not. A sort of surface evil is frequently allowed to triumph, by the way—at least until Hitchcock's blandly disbelieving disclaimer at the end. Serling's playlets may cut more deeply at least in an allegorical or symbolic way, but such implications are usually left in a rather crude state, and the emphasis again goes to suspense and gimmicks.

Both shows are high-grade trash then. But the point is that they have what trash needs (and what much contemporary art does not necessarily have): generally slick, smooth, modishly professional, knowledgeable productions. Hitchcock himself is worthy of more, but his staff produces the series, and he really has little to do with it. On the other hand, I think Serling may really have found his niche in this kind of respectable craftsmanship—at least I find the Rod Serling of "Twilight Zone" a lot easier to take than the more pretentious Serling of "Studio One."

"Way Out," to come bluntly to the point, was not professional. On the few shows that I watched, the scripts were confused and pointless even in their gimmickry. The direction—both histrionic and technical—was often blundering, and even the casting was sometimes incongruous. Some actors seemed to wander around, misreading nearly every speech, and at key moments the cameras might be peering in the most irrelevant directions. The sets attempted realism and looked like random collections of thin properties.

If New York TV cannot compete with Hollywood in producing such diversions, it has no recourse but to produce something really artistic. And if the "Steel Hour" is an example of New York TV's idea of art, there seems little hope that it will. I suppose that in its way, "Naked City" can come as close to artistic honesty as any drama on television, but after all "Naked City" is really a Hollywood production that just happens to be filmed here.

The final irony is that "Twilight Zone" owes a frequently uncomfortable debt to an old New York show, "Tales of Tomorrow," which was never thought of as very much when it was on. And it was on nearly ten years ago.

(*Village Voice*, September 21, 1961)

A Case for the Courts

A couple of local columnists have asked their readers to lament the passing of "The Law and Mr. Jones," a show which stood firmly on the side of democratic right and moral justice, a show on which right and justice were the subject of a great deal of grotesque mugging, smirking, snickering, and facile self-righteousness. The lawyer Jones was acted by James Whitmore, who spent some time at MGM being Dore Schary's idea of Spencer Tracy, you might say. It wasn't so much that for Jones right decisions had nothing to do with moral doubt, or even that Jones was basically a buffoon; I guess that it's just that I don't think I'd like him for a father, and I expect the producers more or less wanted me to.

All of which is worth bringing up only because there is a new show on about lawyers called "The Defenders," and because it is decidedly a different matter—at least its second episode, "Killer Instinct," was. It involved a young stock broker who had apparently been needled into striking a bully, and accidentally killed him. The young man was at first unthinkingly willing to accept his violence as necessary self-defense, but gradually the meaning of his act impressed itself upon him, despite the encouragement of his little son, who saw him as a hero for getting his name in the papers; of his lawyer, who insisted he was not legally responsible; and of a psychiatrist, who explained his blows as a reflex to behavior earlier conditioned by army commando training. All this was carried off with some rather noisy melodramatics and speech-making on the surface, but both in the writing and the acting there was a commendable determination to stick to the moral point and to hint that the moral point is always a personal one and not always an easy one. And obviously the playlet involved an honest effort to face the consequences of its violence.

I am happy to say that "The Defenders" is produced in New York. The episode in question was written by John Vlahos; the idea of the series comes from Reginald Rose; and one of the two lawyers around whom it is built is played by E. G. Marshall. I am sure that whatever their quality, future installments will be interesting at least for Marshall's acting.

Notes: (1) Steve Allen and comic company came back last week as if they had waited impatiently for a year to do it. True, they did drag along Pete Fountain with his quaint imitation of 1938 Benny Goodman, and a personal appeal from Mr. Allen that we should be with "those who think young" and drink that *stuff,* but they are most welcome nevertheless.

(2) In its first episode, "Dr. Kildare" revisited seemed to me to have script and casting problems (which I am not sure are worth discussing), but it also had a remarkably honest and moving piece of acting by Beverly Garland. She was not helped by her tritely written part of a female alcoholic bent on suicide, but she needed no help, and I felt privileged to have seen her.

(Village Voice, October 5, 1961)

And Now for the News: I*

Whether or not there is enough news on TV—and whether or not the capsule form in which most of it is presented has much effect anyway—there is still the question of the pompous "production" claptrap that is usually involved when a TV news announcer sets

---

* My comments on the news shows came at a time when the networks and local stations were managing to expand from the fifteen-minute news format, a hold-over from radio, but before constant criticism from our journalists had virtually forced them to offer more news and documentary shows that raise more issues on more subjects, shows which have attracted advertisers and eventually resulted in one big ratings' success, "60 Minutes." I don't know if radio and TV *can* give us more than the quick snippets of fact, opinion, and feature-story that they gave us then and give us now in such profusion. But I think their writers should give up all references to "details" or use of similar terms, for details are something TV news and even our best TV documentaries *never* give us, and something we should all still seek elsewhere.

out to deliver a ten-minute summary. From the earliest days of television, producers have tried to make such occasions seem important by having their men sit behind ponderously ornate desks, with blotters, pens, and other paraphernalia plainly in view, before backgrounds on which irrelevant maps, books, and standing globes were ostentatiously in evidence. A more recent refinement involves rear-projection screens on which all "on-the-spot" film footage first has to appear behind the newsman before we cut directly to the film image on full screen.

Perhaps the reduction to absurdity of all this is the new James Hagerty version of the 11 o'clock news, where we begin with a picture of a model satellite circling the globe and a puffing announcement about what important things we are going to hear; what important people are going to interpret them for us; and from what important, farflung places. What with the boasting, the constantly announced switching from one broadcaster to another, and the somewhat reactionary editorializing, the program manages to cover about half the ground it should, and cover it with such constantly distracting gimmickry that one is hardly able to listen.

The one public success in TV news programs is of course the Huntley-Brinkley report on NBC. And these two men have, almost by themselves, taken over the leadership which CBS News held for over twenty years. I think Huntley and Brinkley deserve their following. Their news is well selected, their film clips are pertinent and well edited, their reading is forceful but without the intrusion of "personality" ego. But most of all, they stand up in front of a plain backdrop, look directly into the cameras, and tell us what's been going on—no desks, no globes, no gunk—under the assumption that we want to know.

(*Village Voice,* October 19, 1961)

And Now for the News: II

The networks have come to grips in several curious ways with the many recent criticisms that there is not enough news on television. For example, at 10 p. m. on ABC there is one bold headline read between shows, apparently replacing a commercial. CBS has added a 10 o'clock morning news show for housewives, and NBC a Saturday-noon feature for teen-agers.

The housewives are offered a daytime "Calendar," during which Harry Reasoner gets to deliver a few minutes of news summary. Otherwise the program seems to be devoted to converting current events into fairly idle chit-chat, or it becomes a "plug" show on which Reasoner and actress Mary Fickett interview guests who, one way or another, usually have something to sell. A sort of "Today, Jr.," then. But Reasoner can rise to the occasion magnificently, and the dressing down he gave a man who recently brought over a lot of silly gadgets from the "Electra City" show, "appliances of the future," was something to see. ("Automatic scouring device? Well, that ought to clean a couple of pots before it breaks down.")

The teen-agers' show is called, dig this, "Update," and it is presided over by Bob Abernathy, the most curly-headed, baby-faced newscaster you ever did see. The commercials are attempts to get young listeners hooked on the vanity of cosmetics (how's that for bringing them up in the American way of life?). The show professes to help its audience "understand the world around" them—well, we can always blame the world around us when it may be the world within us that has gone awry.

Actually, the "feature" stories on "Update" are some of the best on TV. Example: a recent account of plans for moving ancient Egyptian monuments out of the path of lakes to be formed by the new Aswan Dam.

In a crisis, it is surely a great comfort to us all to know that we can convert world events into a diversion during morning coffee, or feel righteous about sluffing them off on teen-agers.

(*Village Voice*, November 9, 1961)

And Now for the News: III

Many broadcasters who read news summaries on radio and television just don't seem interested in what they're telling us about. Sometimes obviously not. Or sometimes an announcer will cover up by taking a monotonously pompous tone in his words but not toward their meaning. Or he will tell us about everything from earthquakes to hot weather as if he were announcing the greatest event since E Equals MC Squared.

And then there is the TV newsman who grinned pleasantly into the camera as he told of student rioting in Venezuela and the unexpected death of a Nobel scientist.

The really reprehensible type is the man who uses the news as an excuse for some sort of wish-fulfilling "commentary" or personality act of his own. His like used to flourish in radio—Gabriel Heatter was an obvious example, H. V. Kaltenborn a more subtle one—and I expect that, especially with TV reporting under attack, a man who came along with just the right middlebrow, and perhaps Timelike, tone could be an enormous success with columnists and public alike.

On the other hand, there are many, probably sincere, reporters who cannot convey their sincerity to a camera very well. I am sure that I would feel that John K. M. McCaffery was in a terrible rush, even if he weren't (as he usually is). And I always get the impression from Douglas Edwards that everything is going to be all right somehow, and from Walter Cronkite that he is deliberately overstating the seriousness of it all.

Since I find so much television interviewing contemptible, it is probably unfair of me to remark that I find Gabe Pressman's panting encounters inane and often insulting (did you hear him ask Margery Michelmore, late of the Peace Corps, if the moral to her story was never put anything in writing?).

One man who generally speaks to the TV camera with impressive and personal directness is Charles Collingwood, and he often makes "WCBS-TV Views the Press" an exceptional program in the vast wasteland of boring Sunday-afternoon public service and superficial "culture" shows. I was particularly struck recently when he straightforwardly told his audience that ideas do not originate in mass-circulation press but in a limited number of gallant little political and literary magazines. He held up copies of *New Republic, Foreign Affairs, Hudson Review,* etc., and cited examples. Such knowledge should be commonplace, of course, but coming out of a TV set it seemed a bold and daring admission.

(*Village Voice*, November 16, 1961)

Middlebrow Sickcomic

I wish everybody would come off it about Bob Newhart. He is a pleasant and more-than-capable comedian. But surely the fact that his fairly affable jibes at business and politics have been called "penetrating" and "outspoken" is a symptom of something more than the usual carelessness with which the word "satire" is slung around. I wonder whom he is supposed to offend—this rather nice young man whose show features such unspeakably square guests as Roger Williams, who does some of the cheese commercials himself, and who all but falls on his face on occasion thanking his audience for being "just wonderful."

There was one brief sketch in which he fell into a hotel fountain trying to avoid tipping a bellhop; George Gobel or Jack Benny could have done it without a second thought. Another lampoon, supposedly on African nationalism, turned out to be an interview with Tarzan, employing some lightly amusing material on such controversial subjects as tree houses and monkeys. Some happenings in a greeting-card shop might have been made up of lines Jonathan Winters had left over. And "if the builders of the pyramids had had to finance the enterprise like modern construction men" had a couple of good comments, but almost immediately fell into silly and facile anachronisms of modern jive in Ancient Egypt. Steve Allen, Garry Moore, and Art Carney do the same sort of thing almost every time they are on the air, and I'm afraid they do it better. There are sketches on the Perry Como show which cut a lot deeper with a lot less pretentiousness—a recent lampoon of *Peter Pan* and a burlesque involving Como's evening at a night club said some rarely perceived things about the American dream. And simply by being, Groucho Marx can make as penetrating a comment on American mores as our arts are ever likely to offer.

It is no pleasure to report that the Newhart show seems to get less and less funny each week. He was at his early best having private enterprise run a fire department, but a recent sketch about how a supermarket manager hypes his customers was not only not funny, it was rather grim. It is no pleasure to report it, because, as I say, Newhart is a talented and likable comedian. If he does not live up to his notices as a satirist, it is not really his fault because he is not a satirist. What is regrettable is that he is not living up to his own genuine potential to be funny.

(*Village Voice*, November 30, 1961)

What's Up, Doc?

Scripts aside, "Dr. Kildare" continues to be one of the most carefully produced and often skillfully directed shows on television. I hang on to reservations about Richard Chamberlain in the title role, for it seems to me that at crucial moments he doesn't know what else to do but go all-over boyish. If the intention is to have Kildare gradually mature, it probably isn't going to be easy—or, at least, not convincing.

However, there have been remarkably good performances by actresses in several shows. I have mentioned Beverly Garland's despairing alcoholic previously. I also remember Gail Kobe's impressive moment-of-truth in an episode called "Immunity" as a woman discovering that, rather than rising above her background, she was treating her own people with snobbish cruelty. And I remember Suzanne Pleshette's fragile pathos and comedy as a girl, beautifully accepting of her own unchanneled liveliness, suddenly and incongruously stricken by incurable disease.

There is usually a moving scene—really moving, and therefore well-staged and photographed too—in each show. Frankly, I do not often find such calmly self-effacing acting on Broadway, and I seldom find such effective handling of actors and camera in Hollywood movies these days. It is also reassuring to know that television can move us to something besides laughter or obvious revulsion.

Then, if you dig violence, there is the other new hospital show, "Ben Casey," whose surly hero has enough pent-up aggression to suggest advanced paranoia—or at least a good case of what psychologists are fond of calling an "authority problem."

Casey is a success in the ratings, and he is a curious hero indeed. The scripts constantly have him offering arrogant moral pontifications on his fellows, and then applaud him for it. A recent show even implied that such judgment is the function of psychiatry!

Casey has little dignity and—always one-up on even the best of his associates—he will allow little to anyone else. On one show he attempted to revive a failing heart by pounding on its owner's chest—a time-honored medical practice which, at Casey's hands, seemed a gleeful sadistic indulgence.

Note: I have recently remarked that I find most TV interviews depressing; if you saw Steve Allen's fatuous and interminable encounter with Sophia Loren a couple of weeks ago, you will surely know what I mean.

(*Village Voice*, December 21, 1961)

Genteel Wasteland

The most recent inquiry into TV programming by the FCC quickly became an inquiry into the policies of the American Broadcasting Corporation, then became an inquiry into the series called "Bus Stop," and a critique by Newton Minow, the FCC's chairman, of an episode called "A Lion Walks Among Us."

The program in question concerned a baby-faced and morally corrupt teen-ager who apparently tried to seduce an older woman who gave him a lift (and apparently did not succeed, although we weren't sure), who did murder an old storekeeper for a small sum of money, who did burn his own hands after his arrest to gain sympathy against "police brutality," and who convinced his naïvely idealistic lawyer of his innocence. To defend him, the lawyer wrecked the public reputation of the woman by exposing her alcoholism on the witness stand. The boy got off, then murdered the lawyer for the money to leave town. The woman picked him up again, and killed the both of them by driving off a cliff.

If this overloaded plot should mean anything on the face of it, it should mean that we ignore or explain away the continuing fact of human evil at our peril—not an immoral point, and one which some of the great minds of the century have voiced. But a plot as such may not mean much, and handling mean all. And handling could have made this plot into sensationalism, into a suspense thriller, into a bitterly cynical comedy, or into a responsibly compassionate drama.

The handling on that "Bus Stop" episode *was* shallow, but it was not the handling that was under attack. Actually, the handling of every playlet I have seen on "Bus Stop" has been shallow. There was an episode that asked us to applaud when a chronically wayward husband somehow returned to a meaningless marriage and a wife who can never feel anything but superiority toward him when he discovered that his son needed guidance. Moral?

There was another meandering script in which an apparently sympathetic heroine leaves a life which has brought her several husbands and constant promiscuity to try another marriage simply because her own, well-meaning mother wants it. She defies her fiancé's mother to the point where that overbearing old woman has a fatal heart attack. She beds down with just one more old flame after her marriage without much hesitation. When she returns to her husband in the end we are asked to applaud her new insights about her marriage, insights which were decidedly not in evidence. Immoral?

I think so, but it is treatment and handling that makes these smug, unreflecting little stories immoral. *Their* morality would probably never be publicly attacked, any more than the facile moral smugness of the "Robert Young Show" would be attacked. Attack comes when the subject matter itself is more sensational. You know and I know and Newton Minow knows (whether any of us has thought about it) that the FCC is going to be in real trouble if it ever attacks an interpretation, or accuses a TV producer of aesthetic shallowness; it is easier to take one's high-mindedness out on subject matter. It is easier, and it is perhaps a true sign of decadence, for the result can only be that television may have to settle for a smooth, reassuring, and meaningless gentility, which will leave everyone unruffled except the handful of aesthetically honorable men who try to work within it.

(*Village Voice*, March 8, 1962)

The Golden Eggs
of Television

A recent installment of the "Dick Powell Show" concerned itself with the use of illegal "pep pills" among truck drivers, and it produced some grumbling from the local press about television's persistent exploitation of crime. I have the strong feeling that if the same script had been given on the likes of "Studio One" a few years back, it would have been praised as a "strong look at a pressing social problem," or some such.

Which sort of brings us to a recent article in *TV Guide* by Buzz Kulik, who began directing in the so-called "golden age" of television. Mr. Kulik starts out as if he were going to write a version of that annual *TV Guide* piece about the funny goofs that happen on the air on live TV. Pretty soon he is doing a defense of film shows, saying that the idea of the "spontaneity" of the old days is largely a journalist's delusion (Paul Muni, praised for a "pensive" performance, had actually been fumbling his lines), and besides, filming allows invaluable control and polish.

Soon Kulik had introduced his real point: "Live TV was fun and exciting, a great training ground, but it was not necessarily good theater. If all the U. S. Steels, the warmed-over David Susskind and Fred Coe specials, and the last Du Pont I saw are examples of good theater, then I say make mine 'The Defenders' . . . The stories, performances, and productions are as good if not better than anything TV, live or film, has ever had."

Television is vast, he goes on to say, but not all wasteland, and "Somebody's Waiting," with Mickey Rooney; "Three Soldiers," with Telly Savalas; and "Shining Image," on "Dr. Kildare" with Suzanne Pleshette, were plays and performances of plays that any medium should be proud of.

I'm convinced most carping about television is mere snobbery, and Mr. Kulik would apparently agree. "I have a feeling that too many have contempt for the medium. They're all dreaming of theater, of pictures. Been to the theater lately? It's embarrassing. The movies? Well, they have their shortcomings, too."

Monday TV has "Yogi Bear," and "Thriller" (which is either good or pretty silly) or "Ben Casey" (which is either good or pretty sick). Tuesday: "Quick Draw McGraw," the Bilko and Marshall Dillon re-runs, and Garry Moore. Wednesday: the Como variety show and "Naked City" (which is either very good indeed or sick as hell). Thursday: "Huckleberry Hound" and "Dr. Kildare." Friday: Soupy Sales. Saturday, all evening on CBS (which is very good, and currently, in "Gunsmoke" and "Have Gun," very sick—but more of that another time). Sunday: "Bullwinkle" and "Car 54, Where Are You?" The man is right; TV is doing fine.

(*Village Voice*, March 29, 1962)

For Paar

What is Jack Paar really like? How could anyone seriously ask? He is what we have been seeing for five years, a man with a complexity which even his own strong sentimentality cannot simplify. He has succeeded—as has often been said—in being himself. But how many of us realize how much that means, how much it takes for a man to be himself under the pressures of big-time show business?

Those pressures demand that a man take one small, engaging aspect of himself and play that and only that. Remember Joey Bishop in Paar's chair? And Hugh Downs? Were they themselves?

I don't suppose I've ever seen any man reveal his personality and character more fully in public; I know I have seen no public figure so reveal himself in a popular medium—not even Richard Nixon. Fully, and not always wisely. Was it ego that made Paar sometimes act as if no part of himself might offend us?

But what is the point in dwelling on the shortcomings of a

man when even those shortcomings were so obviously and so constantly on display? For me, I think I shall remember longest the night with William F. Buckley, Jr. Mr. Buckley could probably have spent the rest of his life swaggering around the Eastern colleges, his cleverness sure in its dazzling effect. Paar is not afraid to admit he does not understand what he does not understand. And because he is well intentioned, because his intelligence is largely untrained, and because he did not pretend, Paar asked basic, apparently naïve questions. So there was William Buckley, certainly no political thinker, and not even a bright young man, but looking, for the occasion, rather like a smirking, self-centered prig.

(Village Voice, April 19, 1962)

Cultural Interest

Somewhere through the murky darkness it is still possible to discern that "The Untouchables" must once have had something to offer, at least in drive, pace, and atmosphere, on its first season or so. But the grinding monotony of the past two years will convince you that if you have seen a couple of "Untouchables" you've seen them all. Still, there is probably no more succinct comment on the state of American culture than the continuing popularity of this show, as Elliott Ness punches his grim, self-righteous way to victory in the weekly skirmishes without ever really winning the battles, much less the war. Nearly everybody watches "The Untouchables" apparently. Hardly anybody approves of it.

Of course, there is another meaning of the word "culture": what we think we ought to like. It is obviously this meaning which local CBS, NBC, and ABC channels have in mind when they distribute their joint monthly booklet called *Previews* to the TV journalists. *Previews* says it lists "programs of informational, educational, special and cultural interest," and it is obviously designed to convince someone that television is doing the job that everyone says he thinks it ought to be doing.

Have some samples: Regularly featured in *Previews* is "Kukla and Ollie," a sometimes quite fancifully amusing drag show with puppets (you might say), which some adults are convinced is ideal fare for children. And dig this: if the "DuPont Show" dramatizes the Brinks robbery, it's cultural. The sometimes amateurish "Steel Hour" and the bland "Westinghouse Presents" are recommended cultural drama—and about the only TV drama that makes it. The vulgar, mish-mash taste of the "Telephone Hour" provides cultural music. There is also a recommended filmed puppet show called "Davy and Goliath" on which a small boy defies the grand enigma of 5000 years of Western theology and philosophy by "coming to know what God is like." Then, Walt Disney's sticky posturing is cultural but "Huckleberry Hound" is not.

Never mind. Some day "Camera 3" will offer some old Bilko kinescopes the same exalted position it has lately awarded Laurel and Hardy. Bet on it. But it will take a few years.

Commercial notes: Those Kaiser "men of the industry" plugs about the workers—are they selling something, or is there some political movement they want me to join?

Late at night, Radio Free Europe asks you to send in a dollar to protect Ozzie Nelson's world from communism. I'm still thinking it over.

Have you dug Zsa Zsa sniffing the Lestoil? (I thought there was something.)

Then there are those bejeweled females who spray deodorant all over the hostess's john and then tell her they're just fascinated by what she did in there.

(*Village Voice*, May 3, 1962)

Thomas's Muffins

If you watch the "Dick Van Dyke Show" with enough persistence you will find out what the star does—what he does, that is, besides being fairly charming. But mostly you will find a sort of padding which, if it is not exactly true, is certainly tried, and tried several times a week on TV. It is done according to the Danny Thomas formula.

Actually, Thomas' is a refinement of the situation-comedy thing. As radio discovered long ago, if a comedian is going to present himself once a week to a mass audience, he had better get some kind of dodge. He can spend a little time in doing what he presumably does best (that is, in somehow being funny) and spend the rest of the time being merely engaging and hiding behind some presumably diverting and fairly trifling "plot." Thus giving of himself only in snippets, a comic can parley even a small talent into a sustaining career. Thomas's particular refinement involves a small boy (a bit precocious, but not too), a fairly outspoken but decidedly feminine wife, one or two second bananas (for Thomas, Sid Melton and Pat Carrol), befuddled plots with minor crises, and resolutions that are as sentimental as possible. Thomas's company also produces Andy Griffith's show (with Don Knotts) according to pattern, and this season has tried to do the same for Joey Bishop (with Joe Quinn) without success.

Thomas's company does the Van Dyke show with the wife, child, plots, and Morey Amsterdam and Rose Marie. Everyone involved carries the thing off capably, to be sure, and, except when the denouement is a little too sticky, the show is always at least bearable. There are some moments when it is almost brilliant, however, and these come, about one show in five, when Van Dyke does what *he* does, one of his supple, rubbery pantomime bits: When he is a man falling-down drunk who succeeds in hiding it from his wife by a series of sudden hair-breadth recoveries to apparent upright sobriety. Or when he is giving a spirited talk on safety, and bangs, pricks, and slams both of his hands with every drawer, letter-opener, and paperweight on the set, finally making an embarrassed exit with his foot wedged in a waste basket.

For Danny Thomas, I expect the formula does no harm; I expect that anyone who watches him gets all the Danny Thomas there is to get. I wonder, on the other hand, if Joey Bishop could ever be effective without those dead-pan ad libs, and their puckish hostility, which his baby face makes so hilariously ambivalent. Andy Griffith? Well, he was a very good rustic monologist. If the formula now serves him, o. k.; I doubt if it hurts him any.

But I do think it hurts Van Dyke. It gives a man of his potential a way of getting along comfortably, piling up the credits, planning on the residuals, without doing very much. Worse, he may never find out how much really might do. For clearly Dick Van Dyke is potentially one of the most gifted pantomime comedians we have seen in a long time. But as things are going, neither he nor we may ever find out just exactly how gifted.*

(*Village Voice*, June 14, 1962)

The Smoke Is Settling

"Gunsmoke" has been one of the most successful and thereby one of the most imitated programs on TV. It has also been, I think, one

---

* This might be a good spot for a word on that omniscient presence on TV comedy shows, "the laugh track," actual laughs for shows done with "live" audiences (sometimes augmented, "sweetened" as the trade has it) added in from stock tapes of laughter on studio-made shows. Certainly the laugh track is a misused and sometimes obnoxious device. Nothing is more offensive than the canned giggles that accompany inane actions and unfunny lines on some comedy shows. But the real point, surely, is that comedy is best as a shared experience. We have more fun in a theater full of similarly amused and responsive people. We watch melodrama less socially and tragedy, metaphorically, all alone. The laughs that went with radio shows and are used on TV simply acknowledge this psychological-aesthetic reality. And it might be well to remember that, although the laughter on "Lucy" and "Mary Tyler Moore" came from the impulses of a "live" audience, those on the excellent "M.A.S.H." series come from the sound engineers. Class will tell, it would seem.

of the best. This season the series took a chance and expanded from a half-hour to an hour. Things have deteriorated for "Gunsmoke," but only partly because of the change of format.

At a half-hour, the show offered forceful, terse vignettes of character in conflict. Faced with filling an hour, the writers and producers first came up, not with deeper characterization and deeper or more complex action, but with the same sort of vignettes and a lot of padding; initially they not only avoided the challenge, but, in effect, they almost parodied what they had already done so well. Lamentable, for I am sure that the best twenty or so half-hour "Gunsmoke" shows would represent a remarkable achievement.

Despite the somewhat static acting of Jim Arness, Marshal Matt Dillon came across as no crudely conventional Western hero. His superior position and his superior insight carried a superior moral responsibility. He could do wrong as well as right, and he knew it. And he knew he couldn't always be sure which he had done. The show rendered its points in terms of action and character; it did not preach. And it allowed its characters individuality, which means it allowed them moral individuality. It revealed—lo and behold for American dramatic fiction!—that love was giving, not getting or possessing; that the man who has never had a temptation has never been a moral man; that the agreeable, easy-going man whom everyone likes may actually be irresponsible and unprincipled.

Padded vignettes like the first hour-long "Gunsmokes" are not the worst sort of show one could come up with, to be sure. Much more important is the deterioration in the character of Matt Dillon. Time after time this season he has turned on even a casual adversary and shot him dead. There was a time when Marshal Dillon would have had to make peace with himself over any killing, even in self-defense. Lately he has gunned down several men he might have controlled by a quick word or the threat of a drawn gun.

Paladin of "Have Gun" has been going through the same sort of degeneration for almost two seasons now. He has become a brutally self-righteous man, and script after script has him solving all problems by goading an opponent into drawing his gun so that

Paladin, with scornful impatience, can shoot him dead. But perhaps such degeneration is foregone. Unlike the knights of old with whom he begs comparison, and unlike Marshal Dillon, Paladin has only himself to answer to. An inflated ego and a destructive pride may therefore be almost inevitable. Dillon, on the other hand, has a town and its people to protect. A real secular humanitarian, Dillon. Maybe that isn't enough either. Maybe those knights had the right answer after all.

Notes: "The Benefactor," the much-discussed abortion story on "The Defenders," seemed to me too rigged to make its intended point—apparently they did not have time for the good doctor to pat a dog or cuddle a baby. Anyway, the place to begin is with a recognition that the morality of complex men will always have its conflicts —sometimes tragic conflicts—with the laws of the collective community, and if I want to know more about that I will go to Euripides, thank you, and certainly not Ayn Rand. On the other hand, how about writing to the program's regular sponsors, who copped out for the evening, and letting them know what you think of such conduct? They are: Kimberly-Clark (Kleenex, et al.), Brown and Williamson (Viceroy, Raleigh), and Lever Brothers (Swan, et al.). Or you may want to commend Speidel watch bands for stepping in.

As long as I previously quoted Buzz Kulick on the subject, I should say that I think that Mickey Rooney's "Somebody's Waiting" had a ridiculous script of old-time "Studio One" sentimentality, which Rooney's energetic and much-praised sincerity did not salvage.*

(Village Voice, June 5, 1962)

---

* "Gunsmoke" did settle into its hour format of course, and sometimes memorably so. But it solved several of its problems by moving its emphasis away from Matt Dillon and onto its secondary and occasional characters. That shifted emphasis did not confront the increased idealization of Dillon's character, it simply assumed and avoided it. Still, there were memorable shows. The episode with Virginia Gregg, for example, that involved a husband wrongly hanged by a posse and his close-mouthed wife's compulsive revenge.

> "You're a woman going around killing people!"
> "You lynched my husband."
> "But we said we was sorry."

Critics' Choice

Here's a plot: An army sergeant is instructing a group of recruits in the dangerous and exacting business of handling land mines. There is one know-it-all in the crowd who keeps himself and two buddies distracted with smart asides. The sergeant dismisses everyone but the three problem children and tries going through the explanation again, but the wise guy is still at it. So the sergeant challenges him; he calls him forward to show what he has learned. The recruit swaggers up and goes to work; the mine explodes and he is killed.

From this point on the sergeant's life is made miserable by the two buddies, who constantly demand and get special privileges. Soon he feels they are going too far, but when he denies them a pass, the two go to his superiors, claiming this time that he had deliberately caused their friend's death. A new inquiry is set. Desperately feeling that his career is in jeopardy, the sergeant goes to the enlisted men's barracks, engages the two rookies in a fist fight, and makes them admit the truth.

I have been giving an outline of one of the most celebrated TV dramas of the past season, an installment starring Van Heflin on the "Dick Powell" series. The show was nominated for several awards and has been much praised in the press. That praise comes from the same men who chorus constantly about the immorality of television violence and sex, and the paucity of meaningful drama on TV. These people, along with the author, producer, and director of the show, seem to find the sergeant a stalwart and sympathetic hero.

I don't think I need to remark on the sergeant's weakness in

And there was her relief when she was eventually arrested and stopped.

There was a show about a young man who came to Dodge City and was shocked to discover Miss Kitty's profession as a saloon hostess. Eventually, to save him from folly, Kitty has to present herself as a member of an even lesser profession. But to cite some random examples, memorable though they were, is not to do justice to that exceptional series.

granting the two men constant privileges, nor on the way the drama substitutes his condoned violence for its desperate disbelief in a fair-minded judicial inquiry. Let's just take the event itself: What did this unthinking man hope to accomplish by challenging his recruit to handle the unexploded shell? Suppose the boy had handled it properly. Then, of course, the sergeant would not have proved his point. How could the sergeant prove his point? Obviously, only if the shell went off. Which is what happened.

Some drama.

(Village Voice, July 19, 1962)

Ruminations

I sense that the publicity campaign for the establishment of pay television is again under way. I'll make a prediction: pay TV will probably come, and when it comes there will be the same shows there are now, and there will be commercials. Television is *the* popular dramatic medium, as the stage once was and as the movies were more recently, and that fact basically determines what we see on television—not crass producers, not opportunistic network vice presidents, not greedy sponsors. Nothing much about TV will change just because the same millions who now watch for free begin dropping coins in a slot. . . .

Walt Disney, as we all know, is an American institution. You wouldn't get very far accusing him of plugola. For one thing he is highly sentimental, and in America, you know, sentimentality is somehow supposed to confound any such base impulses. But for one example, a recent Disney show was an interminable plug for his amusement park, laced liberally with overt commercials for (a) the sponsor's products and (b) Walt Disney's products. And while Mr. Disney's supposed teen-age friends smugly toured Mr. Disney's park, we were also treated to several gratuitous close-ups of other overdressed teen-agers taking pictures of each other with the sponsor's handy, simple-as-one-two-three camera. Don't expect any investigations, however. Hell, don't even expect any further complaints. . . .

The villain of a book called *TV in America* by Meyer Weinberg is (up to a point, anyway) the sponsor, who determines exactly what goes on television, it seems, and deliberately keeps its quality low, it seems. The book begins with a kind of motto, a quote from Peter Levathes, sometime president of 20th Century-Fox Television: "You've got to look at television realistically, as what it is today. The sponsor buys a show to sell his product. That is the basic purpose of TV. To sell someone's product." Well, it happens that Fox currently has nothing going for it on TV. Perhaps with Mr. Levathes's quote in mind we know the reason. At any rate, Fox did produce "Hong Kong," one of the best examples of good hokey melodrama TV has had—but somehow now I don't exactly feel like thanking either Mr. Levathes or its sponsors, Kaiser Industries, for its quality. . . .

It is always gratifying when a good craftsman takes a popular literary or dramatic form and does something special with it—some of those stories in "The Unvanquished," for example, are delightful for the way William Faulkner has taken the *Saturday Evening Post* formula and turned it upside down. And, well, Sophocles and Shakespeare were doing something of the sort. I think that is why I admire "The Defenders" so much. The point of departure, just as much as on "Perry Mason," is the whodunit. Ah, how well the form is put to different purposes on "The Defenders." I am bothered, however, when producer Reginald Rose issues one of his statements about the series which seem to indicate that he is primarily interested in crusading propaganda and exposé rather than in drama. In the long run I suppose that melodrama, especially limited melodrama like the whodunit, will always be melodrama; it can't become tragedy—that is, it will always be primarily a "plot" rather than a play about human beings. Meanwhile, there are the "Defenders" shows, and the best of them cut much Broadway fare to ribbons. . . .

There was a time when *TV Guide* kept its middle-brow high-mindedness within the brief editorial comments in the front of the book. Now, alas, they are cluttering up the inside listings with a lot of check marks beside "programs of unusual interest." Do you know what is "of unusual interest"? Well (although I bet you could guess), there is Walt Disney, the Telephone Hour, and a documentary on a Girl Scout Jamboree. Take that last one: We have produced a whole class in American society to whom any kind of news report or factual document, Girl Scouts or no, is intrinsically more important than any kind of drama or fiction—to whom, presumably, a knowledge of the fact that the "historical" Jesus would actually have been born in 3 A.D. is more important than a feeling for the parable of the Prodigal Son. I am always a little shocked when I read that TV seeks an audience of a "serious and literate minority of adults" (or some such phrase) for only a few hours of the week, and that it seeks and gets this audience only with—in one form or another—the latest news. However, I suspect that that same audience, or a part of it, also attends showings of old movies at the New Yorker Theatre, and that someday the same people will rediscover "The Defenders," "Bilko," and "Dr. Kildare." Anyway, with our curious sophistication, we still have not killed off the absurd, slightly irreverent, and usually illogical sense of low comic fun that informs "Car 54." And it is good for the soul to know it is still there. . . .

A new series called "It's a Man's World" is worth watching if only for the wonderfully original atmosphere and casual pace which its writers, directors, and actors managed to create—at least on the first couple of shows. It is a very difficult thing to achieve such an individual style for a series, and anyone who feels that all TV has a derivative sameness should look in.

(*Village Voice*, September 27, 1962)

Seasonal Notes

Don't let them kid you. It happens that there are more new Westerns on television this season than medical shows. But there are two of the latter. I have watched "The Nurses" a couple of times. It is hard to believe that some of the same people who are involved in "The Defenders" are responsible for this cheap, histrionic exploitation. . . .

You know about a Roy Rogers adventure show. It's the kind of hokum a parent will think of as harmless if pointless, and hence allow the children to watch. And Roy, he's always on the side of justice and right. Well, on a recent Saturday morning re-run I watched the sheriff announce in the bad guy's barn, "You're coming in with me!" only to hear, "Sheriff, unless you have a warrant, you can just get off my property." With which Roy, in the hayloft above, drew and pointed his gun, saying, "This is all the warrant we'll need!" Yeah, harmless. . . .

Two of the most talented children's performers on television are Captain Kangaroo and Shari Lewis—children, they say, adore the former, and me, I've got a real thing going for the latter. But I wonder if I would let the kiddies watch either one. It's the commercials. Not only do both of them seem to have a squillion of things to sell, but both Captain and Shari do the pitches themselves, exactly as if they were parts of the shows and in just the same child-serious manner with which they talk, dance, and sing. . . .

Speaking of which, have you seen the "Kissie" doll? Push its arms together, and you get a puckery kiss complete with juicy sound effects. Press the button and get affection. My country, 'tis of thee! Only in America! . . .

There are thirty-three new network shows on television this season. I have not yet watched "Roy and Dale's Variety Hour," "Mr. Smith Goes to Washington," "The Jetsons," "Voice of Firestone," "Stoney Burke," "Combat," "McKeever and the Colonel,"

"Going My Way," "The Gallant Men," "The Beverly Hillbillies," "Empire," "The Eleventh Hour," "The Wide Country," or "Andy Williams," and who knows, maybe I won't. I have watched "Loretta Young," "Stump the Stars," "Ensign O'Toole," "McHale's Navy," "I'm Dickens, He's Fenster," and "Don't Call Me Charlie," just about once each, and who knows, maybe that will be it. I keep meaning to watch Stanley Holloway on "Our Man Higgins," but I haven't gotten to it yet.

I watched "Fair Exchange" once. Eddie Foy, Jr. is not a great comedian, but he is very good, and it is always a pleasure to see professional skill like his. But he is almost the only one connected with the show (writers, producers, actors, directors) who seems to realize that comedy has its own slightly unreal perspective. Foy aside, we watch a group of smug, indulgent, selfish, or stupid people presented as if they were real—which means that they are not so much funny as they are unpleasant and sometimes repulsive. All of which is worth bringing up only because so much TV situation comedy makes exactly the same mistake of viewpoint and style. . . .

"Saints and Sinners" seems to me pretty cheap stuff. The night that hot-headed upstart Nick Adams saved a wise and mature judge, played by Lew Ayres, from impulsively choking his son-in-law, I was a little too upset to laugh. And the new Lloyd Bridges show— very hasty and confused scripts mostly. And I include the celebrated "A Pair of Boots" episode, publicized as finally having been filmed after its author battled the Philistines along producers' row and Madison Avenue who wanted to change his script. Well, there was a great deal of yelling and jumping and shooting and gnashing of teeth, not to say pretentiousness. (Sometimes I think those who insist on doing Something Good on Television are the worst of all.) . . .

(*Village Voice,* December 13, 1962)

How To Fail by Succeeding

If I tell you that "It's a Man's World" was about life in a small American college town, I may very well conjure up all those ludicrous but strangely cherished memories of college musicals, winning the big game, and all that. If I tell you further that it was also about a group of young people, ages roughly fourteen to twenty-two, I will probably conjure up memories that are more immediate and more frightening of a dozen self-satisfied and unreal teen-agers in as many TV situation comedies. And that is lamentable, for by and large "It's a Man's World" has tried to find its own way. And, we should all know, it takes guts to do that in any art or any entertainment. It has been, heaven knows, an uneven series, but it has tried with an uncommon effort to present us with the reckless energy, the time-wasting idleness and boredom, the friendliness, the sexual awakenings, and the loneliness of young people who are not rosy clichés.

The series has also made a most interesting and frequently effective effort to discover just exactly how many cinematic techniques will come across on filmed television, and that is a job that certainly needs to be done.

Peter Tewksbury, who conceived and produced and often wrote and directed, was, surprisingly, associated with "Father Knows Best." It is less surprising that he was also associated with the early episodes of "My Three Sons," for some of the style of "Man's World" was tried out there, and, on one or two shows, much of its abruptly integrated use of music—but little of its whimsy and almost none of its forcefulness.

I will need to say something about the characters and situation of the series, but by doing so I will betray one of the most engaging aspects of its presentation. For there has been no compulsive exposition; we have been allowed to meet and get to know the leading and supporting characters casually and gradually, and it took me nearly half a dozen episodes to find out what I will tell you now.

The scene is a small Ohio river town called Cordella. Two of our four main characters attend the college, and all four live in a houseboat. Wes is a local boy, in his early twenties, a pre-law student supporting himself. He is the guardian of his fourteen-year-old brother Howie—they have apparently been orphaned recently. Wes is sober, orderly, sometimes overbearingly so, and he is engaged to a local girl, Irene. Tom, called Tom-Tom, is an upper-class Chicago boy, defiantly going to school in Cordella by his own choice, the humorist of the group who occasionally strikes an intellectual pose, but at this stage much involved in reacting against the standards of his parents. Somehow these three have taken in Vern, a quiet youngster from North Carolina on his way to Hollywood—he has decided to make a lot of money as a country and western singer. (If all this sounds somewhat bizarre, I can promise you that none of it was ever treated as if it were particularly bizarre.) Each of the boys has formative individuality, and each holds the others in a real but slightly embarrassed affection.

It does not surprise me that "It's a Man's World" has apparently received little encouragement from members of the higher echelon at NBC (they have decided it will go off the air after January 18). But I am foolish enough to allow myself a little dismay that the TV journalists—at least those I read—have treated it with an airy contempt more than once. (That also means, I note, that they have watched it more than once, and they might ask themselves why they come back if they really dislike it.) We will never, I submit, get better television by the sort of thing that most commentators do recommend: setting up cameras in front of a classic piece written for the theater.

One might say that in adopting cinematic techniques to his show, Tewksbury is simply using the aesthetic discoveries of D. W. Griffith, but he is using them in nothing like the derivative and complacent manner that "Play of the Week" used Broadway's.

"It's a Man's World" was photographed and edited with care, exceptional care for television. I remember several marvelously constructed sequences, low-angled and rapidly montaged, that captured the noisy hubbub of youth on a spree in both its excitement and its sometime pointlessness. I remember a shot of Wes waking up in his bunk cross-cut with one of his girl waking up in her bedroom in such a way as to suggest they were in bed together but also to let us know they were not and that they would like to be and that they knew they must not be and that they loved each other very much. Nor was it an isolated stunt, this double scene, for it was pivotally expressive of the point of the episode, which was their waiting relationship. I remember another wonderfully staged scene in the houseboat, with all four occupants at odds with each other and hotly fighting it out as they cleaned up the place, a procedure which took a delicate cooperation in manipulating brooms, holding dustpans, and the rest. And there was the striking scene that had Tom-Tom photographed against the night sky as he danced around the top of the houseboat, loudly reciting "Hooray for Pooh! But what did he do?" and secretly trying to decide if it were true, as everyone had been telling him, that he had selfishly idled away his day. Superimposed on Tom and his dance were rapid images reprising the events of his day: his impulsive ride on the back of a garbage wagon, his mock gun-battle with two little boys playing cowboy, his odd chance meeting with his girlfriend who was somehow reading a book of mathematical theory in the Cordella cemetery (I know this sounds affected, but it wasn't), and his fine jaunt down the main street, leaving behind the library books that "absolutely had to be returned today or else."

The best episode that I have seen was, except for one slightly sticky scene, something of a minor comic masterpiece. It was called "Drive Over to Exeter," Exeter being the nearby sin town. All good comedy depends on certain more or less familiar situations. They mean little in themselves; the freshness with which they are handled means everything. But I will risk the familiarity long enough to say that the situation here was that Vern announced after a couple of moody days that he thought he would go over to Exeter, and Wes and Tom-Tom assumed he was going to, well, lose his virginity. From this point, all the changes were delightfully rung, from vague philosophizing by Wes about "growing up" through nervous lectures from Tom about lust and a series of acute fallings-out among Wes, Tom, and their girlfriends over just who knows what about what goes on at Exeter, and how. (Meanwhile, there have been hints that what Vern actually wanted was a tattoo.) And at the end, there was a brilliant scene with Vern returning in the middle of the night to explain, at great length and in a state of near-exaltation, to his hopelessly sleep-drugged roommates that he found out he had actually been homesick and had telephoned Mama and Daddy. Then, in a genuinely sympathetic representation of small-town Southern speech, complete with whoops, Vern excitedly gave this logy and grimly uninterested crew a run-down on just what Mama said to him and he said to Daddy and the latest about Aunt Lou.

For a while there was an effort to save "It's a Man's World," cut to a half-hour. This might have been a good idea, for sometimes the effort to portray the leisurely pace of youth led to slow going rather than the desired effect. At any rate, they did enough episodes, one hopes, for rerun on local channels. Meanwhile the networks give us Dobie Gillis, Ricky and David, and "Don't Call Me Charlie." Do they really have a bigger audience than watched "It's a Man's World"?

(*Village Voice,* January 17, 1963)

Beyond the Real City

Among the shows threatened with cancellation after this season is "Naked City," and guardians of past culture—be they museums on 53rd Street or Channel 13—might prepare themselves to revive the best episodes. To put it bluntly, "Naked City" has probably had more successful installments than any other dramatic series there has ever been on television. And that means that it offers some of the best drama, some of the best acting, and some of the best direction seen anywhere—Broadway, movies, anywhere. Anyone who has watched four "Naked City" episodes has seen at least one very good one; anyone who has seen five has probably seen an excellent one; and there have been over one hundred of them since the series converted to an hour and hit its stride three years ago.

If a "Naked City" show fails, it fails of a kind of neurotic or brutal exploitation of its subject: the one about a psychotic painter who turned murderer was disturbingly gratuitous. Or it fails out of a pat pretentiousness: recently we were promised a look into a man's soul as detective Adam Flint attended the execution of a murderer he had helped arrest. Actor Paul Burke's face was rather blank; perhaps he was looking into his soul, but we never knew what he found there.

The series began as cops and robbers melodrama, and the point of departure in most "Naked City" scripts is still apt to be bizarre. A boy whose face was scarred in infancy is living wild, prowling the upper Bronx at night. Or the theft of valuables from upper East Side apartments is being masterminded by the apparently well-behaved son of one of the well-heeled residents.

The second premise of "Naked City" is the reality of location production in New York streets and buildings.

With a good script and a good director, this combination of the exceptional and the literal yields a dreamlike quality. But not the vagueness of Hollywood wish-fulfillment; rather the vividness and immediacy of mythic visions. A bleak upper West Side street, where enraged Puerto Ricans are hurling bricks at cops and passers-by, has the despair of a landscape in a contemporary hell. An empty downtown warehouse becomes, by lighting and camera work, a cavern in an underworld. Neither place has the transience of mere social documentation; both, we realize, are deeper psychological places where each of us may encounter his neighbor, or himself. The paradox of the bizarre and the literal, I think, gives "Naked City" moments of higher perspective, beyond realism—as surely as Pirandello, Ionesco, and Beckett have offered theatrical styles beyond realism. But whereas our contemporary playwrights consciously turn to the theatrical or vaudeville traditions for help, "Naked City" has largely found its own intuitive way.

Sometimes "Naked City" is almost totally successful. The recent story of a police informer, beautifully acted by Frank Gorshin, found convincing pathos in the plight of a nether-world being and managed to teach one of the series' regulars something about his own moral complacency. On another show the proddings of a vengeance-ridden young cop nearly destroyed a chief of detectives, but in the long run helped that chief affirm his convictions. Indeed, I think the most successful scripts believably discover the redeeming side of despised men—and the shadow side of heroes.

In almost any episode there is apt to be a splendid scene: I shall never forget the lonely tenement housewife, in her moment of glory, acting out—deliberately, enthusiastically—the argument she had heard between the pair who lived downstairs. Nor shall I forget a brilliantly staged Village barroom poetry contest between an aging alcoholic writer and a young upstart, which ended as the younger man smugly revealed to the swirling crowd that the elder was not improvising but reciting another's verses.

Any series seems to have a life of its own, and even the best series must sooner or later decline. I do not know if another season would bring the decline of "Naked City" (this season certainly got off to a shaky start), but I do know that one is privileged to see those whom one respects even in decline. Another season of even a declining "Naked City" would be that sort of high privilege.

(*Village Voice*, March 14, 1963)

The Coasters

It has become reflex action on the part of certain television reviewers, whenever the medium is attacked as derivative or unproductive, to point with pride at Sid Caesar and Jackie Gleason, two comedians who had been around for some time before they discovered television and television discovered them. One might expect that there is more than mere coincidence involved; that is, that each man had particular qualities which the television screen itself brought out—that if there had been no TV, Caesar might still be working the Borscht Belt and Gleason getting an occasional part as comic relief in Hollywood films.

Frankly, I wonder. And the return this season of each man to regular broadcasting has done nothing but increase my doubt.

Jackie Gleason seems a very likable man, and that likability registers beautifully on TV. But likability, plus some hoary but venerable jokes, are just about all he has been offering us lately. Otherwise Gleason plays adroitly sympathetic straight man to Frank Fontaine (and if he is new to you, where have you been for the past fifteen years?), or to Rip Taylor, or to Alice Ghostley (in a questionable exploitation of a hopeful spinster). Or he gets out of the way while a nightclub perennial like Henny Youngman does one of his routines.

Not that Gleason started the season this way. I fondly remember a good pantomime about husband and wife waking up of a bleary-eyed morning and constantly missing each other in cross-purpose trafficking across the kitchen; I remember the welcome return of Reggie Van Gleason and the Mother Fletcher salesman ("And now back to our late movie, 'The Werewolf of Trenton'"); and I remember a superb burlesque ballet in which Gleason managed to parody almost every classic movement, while doing little more than tripping and stopping around the stage with his arms at half mast, his hands flapping slightly, and his head occasionally thrust from one side to the other.

The longer the season goes on, however, the more reluctant Gleason seems to be to affirm his reputation, the more willing merely to ride on it. If you're going to coast, don't coast on mere likability. Get a funny way to coast. Like Jack Benny, who has made a career of it.*

As for Caesar, well—I have as fond memories of the old "Show of Shows" as the next man, but aside from a few moments of comic harassment, those memories don't always center on Sid Caesar. It seems to me he is far from being the satirist he is usually called, because he has much too much of the low comic's penchant for pulling anything, absolutely anything, for a quick laugh. I remember a lampoon of the film *A Place in the Sun.* Caesar rolled his eyes and twisted his mouth and delivered such asides as "What is with this chick?" some of which were funny, but none of which had much to do with the subject at hand. Imogene Coca, however, confined herself to a disciplined, gentle, keenly observed parody of Shelley Winters's fine petulance in that film.

Otherwise, Sid Caesar frequently seems to be fumbling: fum-

---

* I should point out that these remarks apply to Gleason's first, tentative season with his "American Scene Magazine." I would not, in retrospect, apply the remarks to his subsequent seasons (through the spring of 1966), and certainly not to his re-alliance with the superb Art Carney for a revived "Honeymooners" which continued for three seasons more.

A footnote is surely not the best place for praise for Art Carney, but for now it will do. When Peter Sellers died in 1980, I was provoked by a friend into reflecting on whether we have a comic actor of comparable flexibility and resourcefulness and potential range. We do of course in Carney, but I despair that we will ever offer him the almost reverential respect we are apt to award a Briton. A part of the reason is of course that Sellers worked, however peripherally, within a theatrical tradition four centuries old, and an unusual stability and respect accrue to any outstanding performer who does. We do not have such a national tradition of course, but if we did I am convinced that Carney would be offered the roles, the special vehicles, and be awarded the reputation his talent and skill deserve. We do not have that tradition, no. But the first step in building stability may come about when our best critics, intellectuals, and academics begin to take a more responsible attitude toward our native artists and our native arts.

bling with a piece of possibly good material that he never gets to, fumbling with a comic point of view he hasn't clarified or a pose he hasn't crystalized, fumbling with lines or blocking he isn't up on. And in his current TV appearances, Caesar seems to be fumbling in shows that are perhaps partly rehearsed, never timed, and conceived without a climax or a conclusion—even without a point. Yes, and fumbling with a sizable talent he hasn't really come to terms with.

His work is spontaneous, you say? Well, with the kind of improvising that the "Premise" or the "Second City" people can do (not to mention Mike Nichols and Elaine May), who needs this?

(*Village Voice*, April 18, 1963)

Memos

James B. Fisher wrote to the *Saturday Review* to complain of the "parlor intellectualism" and "endless talk" on the air that is "especially dismaying to those of us who prefer the printed page as our source of intellectual stimulation." Let him know I agree . . .

Somehow try to get through to the more venerable critics of our culture with the feeling that Walt Disney's reputation is perhaps a curious hold-over, or that he hasn't done anything really interesting for over twenty years (say, since Donald quit being a furious little duck and became a cutie-pie). Try to explain that his successors like Chuck Jones at Warner Brothers (Bugs, Sylvester and others) and several hands at MGM (Tom and Jerry) did wonderful cartoon comedies that make much of Disney's early work seem a crude beginning. Disney seems to be able to get away with almost anything. Not just the constant plugs, but the fact that a large measure of his programming is old theatrical films. Everyone laments Monday, Saturday, and Sunday Night at the Movies, why not this? . . .

Several psychological and analytic societies are expressing alarm at the miraculous thirty-minute cures wrought on "Eleventh Hour." Better they should worry that the cures are wrought by cold-eyed bullying on the part of Wendell Corey . . .

This audience-courting "Johnny Carson" who is on TV now, is he the same Johnny Carson who ran an impudently funny show from Hollywood a few years ago and who took over "Tonight" from time to time from Steve Allen with equally impudent wit? Sure, he *looks* like the same fellow . . .

Ask somebody how it is that when a former network show re-runs on local channels, it carries more commercials and more commercial time . . .

Try to puzzle out why even a good drama seldom looks good on tape, but a variety show (Eddie Cantor's old one, for instance) will always look bad on film. There is a key to something important there . . .

Ernest Flatt is choreographer for Garry Moore's variety show, and Ernest Flatt does consistently good work, week after week. Sometimes he is far better than good—there was the marvelous abstraction of hoedown and square dance steps he devised for four male dancers on the recent show Moore taped in Las Vegas. It made all the Broadway re-workings of Agnes DeMille's "Rodeo" (done by Miss De Mille herself as well as imitators) look sad indeed. Many superb and imaginative craftsmen like Flatt give their energies to television. By and large, all the thanks they get is the thanks that a few insiders can give them, and they can give themselves. Perhaps that is the best kind of thanks anyway. . . .

The "Festival of Performing Arts" was as uneven as such innately middlebrow undertakings are apt to be. I confess I can make nothing of Zero Mostel's word-at-a-time readings of Joyce and Shakespeare. But his roaring impression of a harassed airliner, prevented from landing at one city after another by bad weather, showed fine comic perception. Joyce Grenfell was simply superb. She can move us to tears with so basically banal a situation as a dedicated spinster's cherished weekend visits from her brothers' children, and two minutes later she can double us up with sympathetic laughter at the observations of her cockney character, Shirl, who, as she says, "just talks." . . .

Someone correct me if I'm wrong, but it strikes me that the recent controversy at Channel 13, and whether to make it our "educational" channel, boils down to whether the station is to present a *Herald Tribune* idea of culture or a Teachers College idea of culture. Anybody got a coin? . . .

Bob Hope has won another award. Ay, yes, Bob Hope. The comedian as smart-alec; the comedian as wise guy with a slight smirk. If Hope arrive, can the frenzied paranoia of Lenny Bruce be far behind? . . .

Howard K. Smith posed "What's Wrong with Hollywood?" and answered that the future lay, at least in part, in independent, inexpensive, but artistically commendable productions like Frank Perry's *David and Lisa*. Ho hum. It is a solution we have heard posed all our lives. Careers from Josef von Sternberg's in the twenties through Stanley Kubrick's in the fifties have begun with such auspicious modesty. But no career has ever been sustained by it. Val Lewton is the only man who ever dared try to sustain his without resorting to some form of movie opulence . . .

The "Dick Van Dyke Show" has progressed by leaps this season. Not only is Van Dyke doing more pantomime, but producer-writer Carl Reiner has polished other elements of the show to keenness. The comedy operates on nearly everybody's level, from the lowbrow jokes of Morey Amsterdam through the wise-cracking female of Rose Marie to the more sophisticated exchanges of Van Dyke and Mary Tyler Moore. Better still, Reiner has turned the Danny Thomas-domestic situation formula inside out, so that much of the fun comes from a kind of sly parody of its sentimental clichés . . .

(*Village Voice*, May 23, 1963)

On with the New

I listened to the new chairman of the Federal Communications Commission, E. William Henry (age thirty-four, undergraduate at Yale, law at Vanderbilt), sitting for a verbal "Portrait" with questions from Harry Reasoner, waiting for him to make some aesthetic pronouncement about television. On the whole it was rather a long and disappointing wait. He seemed to be bringing an able lawyer's mind, plus seven months' experience with the FCC, to his job, and he seemed to know all the rules and regulations about what the Commission can do and can't do and is supposed to do.

True, he said that he had always been interested in television and liked it. And he said that he would like to see more variety (the current jargon is "diversity") in programming during prime evening hours. (And who wouldn't on the face of it?) He also said that while Newton Minow was making the pronouncements, TV offered three times as much public affairs programming, and that everyone would agree that this is an improvement. (Yes, everyone would, if everyone would agree that any news broadcast is better than any drama—but everyone wouldn't.) Underlining his point, he said he would like to see television a "social force" rather than just a "medium of entertainment," whatever all that means.

He spoke with strong implicit approval of a CBS production of *Hamlet,* a play which many a right-thinking Elizabethan would have considered "just entertainment," vulgar at that, and which, one would think, has much too much violence and sex (not to mention morbidity) for television.

Finally, there was a lot of palaver about "elevating the public taste" with "high-level programming." Well, I hope they don't overdo it, because if the public's taste ever gets so elevated that it no longer watches the likes of Dick Van Dyke or "Naked City" or Carol Burnett or E. G. Marshall or Bullwinkle or Red Skelton, then television, as well as the public taste, is going to be in a lot of trouble.

(*Village Voice,* June 20, 1963)

Return of a Legend

Under the vaguely arty title "Reckoning" CBS is re-running plays from television's slightly distant past: half of them are being drawn from the old "Climax!" series (the punctuation, I assure you, is theirs) and the rest come from "Studio One." So "Reckoning" gives anyone interested a chance to check on a legend, because "Studio One" in the minds of many was one of those early hour "live" shows that offered high drama, television at an artistic peak from which it has declined into trite formula stuff, canned by Hollywood.

It seems to me that it would be hard to imagine playlets more frequently or more rigidly in formula than those on "Studio One," or a formula more complacent; "Ticket to Tahiti," the first "Reckoning" re-run, illustrated it handily. Briefly, the plot presented a mild, middle-aged widower who unknowingly indulges his wastrel son and daughter-in-law. The widower's fiancée sees through this blindness and in a crisis scene strongly presents him with it. Whereupon he revises his attitude, gives his offspring some stern guidance, and gives up his own pipe dream of an extended vacation in Tahiti in the bargain.

There we also have the formula: a protagonist with a fault who is confronted with it, usually as exposed by a sympathetic friend, and is immediately revealed as a changed man. It's that easy. I won't say he has a "flaw," although I have no doubt that many people associated with this series would have called it a flaw, even a tragic flaw. And sometimes he had only an annoying habit presented, in strident scenes full of sweaty closeups, as a deep and significant problem.

Well, they all have deep and significant problems, these protagonists, so deep that all they need do to deal with them is have a
heart-to-heart listen with a well-meaning friend.

And from such naïve anti-tragedy has come the story of TV's
glorious past. No, naïve is too mild a word. For working purposes,
let's call it decadent.

(Village Voice, July 25, 1963)

Summer Sequence

If anyone is tempted to believe that breaking up the big networks and
establishing a sort of community-theater-cum-journalism through
local television would bring an improvement in the medium—a position bluntly asserted by Paul Goodman in a recent *New Republic*—
he should take a look at "Repertory Workshop," a series of taped
dramas done by CBS affiliates. I confess I have seen only a couple
of these programs all the way through because I could bring myself
to watch no more. One (which shall be nameless) persented a repu-
table American one-act play with the strident naïveté of little theater
at its worst. The other was a dramatization of Katherine Brush's
short story "Nightclub," chicly retitled "Powder Room," taped in
Los Angeles. The amateurishness of the actresses involved was of
the sort which a good Hollywood director instinctively disguises.
The director of "Powder Room" took an opposite approach and
enhanced every flaw in his production: lighting, camera position,
cutting, pace, and performance. . . .

The scattered new programs produced during the summer take
almost as much journalistic abuse as the preponderant re-runs. But
one summer replacement that has been welcomed is a musical series
called "The Lively Ones." It seems to be the opinion of its producer,
Barry Shear, that there is something wrong with presenting a musical performance photographed on a stage. He therefore once offered

Dizzy Gillespie performing in a ball park. (Get it: Way out in left field? How's that for yesterday's hipness?) It is the opinion of this reviewer that Dizzy Gillespie is worth (a) hearing, and (b) watching closely for his mental concentration and physical involvement as he improvises. And that it does not matter much where one places him. The result of Mr. Shear's approach is a kind of *Vogue*'s eye view of vaudeville. And I have the feeling that Mr. Shear's former associate, Ernie Kovacs, would punctuate this manic *chichi* in a split second. . . .

A recent discussion on the "Open Mind" asked: "Is the cultural boom a bust?" And do you know who did the discussing? Well, there was William Saroyan and David Merrick and Jean Dalrymple and Dave Brubeck. And there you have the fundamental reason-at-the-source that most television culture and public affairs programs are simply not worth watching. . . .

Both the "Premise" and the "Second City" companies have had extended local TV exposure recently. Not too surprisingly, "The Premise," which comes off stirringly in person, came off less well on TV. The sketches would have been better served by better staging and camera work, and by being less often fatuously interrupted by David Susskind. But the real consideration is that "The Premise" depends on getting its audience rather directly involved. "Second City" plays *at* its auditors and thereby registers a bit better to the camera. The nadir of photographed theater was the apparently good Central Park production of *Antony and Cleopatra* as taped by WCBS. Performances broadly pitched to a large amphitheatre looked coldly ludicrous when watched by nearby TV lenses. Culture, indeed.

(*Village Voice*, August 22, 1963)

New Season: I

You can just see them pasting together that "Arrest and Trial" show. "Ninety-minute shows are gonna be big next season." "Look, cops and crooks are always good if you get a fresh angle." "How about all the lawyer shows? Can't we do something with that?" "Don't forget the head-shrinker programs. There'll be two of them next season. Maybe we could throw in a little psychology."

The origins, the derivativeness, the angles, even the glue and the padding—they are show. Yet "Arrest and Trial" is not a bad show at all. Ben Gazzara has so far been willing to act. Anthony Franciosa was willing to in the first episode. (Chuck Connors can't act, but he's willing to try.) The pace and plotting—crucial matters in melodrama—are the work of professionals. The padding is good; that is, it is diverting. The bargain-basement Freud is offensive—but so is the bargain-basement Freud in middle-period Eugene O'Neill. "Arrest and Trial" will probably survive and, such as they are, have pleasures of its own to offer . . .

From all indications, Imogene Coca's "Grindl" is not going to be very good. And I am especially sad that this very subtly talented woman should be involved in another failure. I plead with someone to consider the possibility that Miss Coca is not a lovable fey gamin, all thumbs—that beneath that perky face there is a determined brain, a shrewd perception, all claws if the occasion arises. The woman is just too complex to be merely cute, and just too shrewd to be blundering. . . .

There is no question that the handling of the "Negro revolt" (the quotes are mine because I don't like the expression, but admit we're stuck with it) and of the Washington March by various network news departments, particularly NBC's, has been very good. But I wonder if the handling, and the subsequent self-congratulatory full-page ads in the *Times,* are really evidence of a "new maturity."

It might be a surer sign of maturity if such reportage on television were accepted as a matter of course, as it should be. . . .

From their beginnings, both the "Premise" and the "Second City" have been able to gain appearances on every "plug" show, network and local, on television. And each has had a full performance on TV when a given show closed on stage. Not so the integrated "Living Premise," which has gained only a couple of radio plugs for itself and is about to go north to the CBC in Montreal for its end-of-run television reprise. Now, just guess why. Yeah, that's it. And furthermore, the refusals were blunt and specific. And they came in this summer of 1963.

(*Village Voice*, September 26, 1963)

New Season: II

That five-part, name-ridden opening to the new, Zimbalist-only version of "77 Sunset Strip" stretched an old, half-hour private-eye radio plot farther than TV has ever dared stretch one before. As for the miscegenation story that followed, it made a rare effort to use character and social conflict as the ultimate keys to its puzzle.

On "Burke's Law," I have rather fond memories that Dick Powell played its lead character, a suave police detective who is a millionaire in private life, with a fair degree of impatient charm a couple of years ago. But Gene Barry. And those scripts with what passes for sophisticated repartee. Well, try to imagine Leo Gorcey's idea of William Powell. . . .

Subsequent episodes of "Arrest and Trial" strongly suggest that I might be wanting to take back all I said about the series previously. . . .

Jerry Lewis has been given a prime-time desk-and-couch show by ABC. Lewis is a very talented man, but I am not convinced that he has done the best by his talent. Informal, live television has worthwhile qualities, but I am not convinced that Jerry Lewis is any kind of conventionalist, nor that he knows how to discipline his ego on informal, live television. . . .

The new Phil Silvers series is depressing. You can count the comedian's best Bilko shows among the classic TV comedies, but now he is involved in the kind of bad West Coast imitation that would be unworthy of a second-stringer. . . .

You probably will never hear another word from me about "Glynis," "The Greatest Show on Earth," "Harry's Girls," "The Farmer's Daughter," "Temple Houston," "The Patty Duke Show," and a couple of others. And you may be just as well off. . . .

I defy anyone to watch five consecutive toy commercials (any five) and not come away wondering about the state of the nation. . . .

Danny Kaye has so far been getting by on charm, pleasantly diverting routines, guests, songs, and at least one really hilarious moment per show. If he can keep it up, he should last forever, regardless of ratings. (Did you catch his one-sentence parody of all cigarette commercials? "Milder . . . because we say so.") . . .

(*Village Voice*, November 14, 1963)

One Man's Education
(*Another Man's Hokum*)

Early this fall, a series of ads for the Blue Angel supper club in New York (relatively small, a little more than relatively chic, and absolutely located on East 55th Street) announced the arrival of Max Morath, "the sensational singing ragtime man." Morath, the ad also revealed, came complete with praise from Jack Gould, television critic of the *New York Times* ("presides over a wonderful rag piano—and lets go") and from John Chapman, playgoer for the

New York Daily News ("heard him in a honky-tonk joint in Cripple Creek, Colorado, and have since become devoted to him").

Morath's booking was also preceded by the issue of an Epic Records LP (LN 24066), not his first record, but his first for a major company. His complete act apparently includes lantern slides and a little more direct audience participation than would make sense on an LP, but the essence of what he does was recorded. He wrote his own liner notes, too, and he has this to say about the "Maple Leaf Rag": "Here, unadorned is Scott Joplin's first ragtime classic from 1899. More than any other piece, it triggered the ragtime revolution.

"Ragtime is probably the most misunderstood music in history. As one critic has said, it was truly the 'folk music of the American city.' But in a few years it became 'the old, old story of Tin Pan Alley watering, nay drenching some exceedingly valuable musical material.' Honky-tonk tricks soon crowded out the true ragtime of the storied Negro players who originated it."

"Maple Leaf" is indeed the work of a gifted man and a serious popular artist (there is no contradiction there). So are "Dill Pickles," by Charles Johnson, and "Ragtime Nightingale," by Joseph Lamb. Morath performs them all with faithful respect on his Epic recital. Also, alas, he performs with a reverence that approaches lifelessness.

But Morath is not a lifeless performer. He is not lifeless when he is doing "I've Got a Ragtime Dog and a Ragtime Cat . . . A Ragtime Wife. I'm Certainly Living a Ragtime Life" on his jangling, "prepared" piano.

To put it another way, Max Morath's tastes, inclinations, and interests are more inclined toward "Tin Pan Alley watering, nay drenching." And his stilted tribute to Scott Joplin seems rather deflated when placed beside "Where Did Robinson Crusoe Go With Friday on Saturday Night?" which he really tears into. True, his knowledge of American popular music circa 1890 to 1918 does seem broad, one hastens to add, if not deep.

Certainly, it is impossible to take offense at Morath. He is a good, engaging showman, and an entirely honest one. Indeed, he would be engaging if only because he obviously enjoys so much just being "on." And he loves hokum and nostalgia. He loves it, also, with none of the patronizing or hints of sordidness we are all too used to from every side, from *Sammy's Bowery Follies* through TV variety numbers.

Max Morath has conducted—derby hat, garters on sleeves, showbiz savvy, and all—not one, but two series for National Educational Television: "The Ragtime Era" and "Turn of the Century." That Morath can be entertaining is self-evident; that any of this stuff is *educational* is damned near preposterous.

We have seen both these series in New York on Channel 13. Lucky us! We got *educational* TV! But "The Ragtime Era" ran on a local commercial channel before we had such a thing as educational TV. And a local commercial channel is where it belongs. This is commercial entertainment (and good of its type), but some of us think an "educational" channel should set its sights differently, if not higher.

(*Listen*, January, 1963)

All Around the Town

"East Side/West Side," a CBS series about social workers, met with an initial warmth from reviewers which had since cooled. A recent tale about a compulsively gambling cabbie, by taxi driver-writer Edward Adler, keenly and realistically observed its milieu. And "Who Do You Kill?" by Arnold Pearl was as forceful a piece of social propaganda as has been on TV for a long time.

Pearl's playlet concerned a Harlem couple. The wife hopefully hustles drinks to support the family; the husband half-heartedly takes courses toward a vaguely promised job and minds their new baby. During a family argument, a sudden scream from the next room indicates the baby is being gnawed by a rat—a not uncommon occurrence in some New York slums. The father takes the bleeding infant in his arms and has to walk to a hospital because no cabs will stop for him.

Who is to blame? Who do you kill in revenge? The City had served the landlord with papers; the landlord had started anti-rodent repairs in the building; the doctors did all they could. The frustration and red tape swirled around a dying child.

All of this was presented with a glaring and sustained urgency, and with a production which captured the callousness of big city life down to the slightest details—like the empty faces that surrounded the wife in the manic, juke box-impacted atmosphere of the neighborhood bar. And the show was helped enormously by the acquisition of cameraman Jack Priestly, late of "Naked City."

But "East Side/West Side" has not often achieved such pointed documentation. Most of the time, whatever the subject of the show (urban renewal, public housing, illegitimacy, block busting, retarded children, delinquency), we have been offered a vague and slightly rigged debate on subjects whose essentials have eluded the participants. Sometimes these debates trail off with an arty inconclusiveness, and at least one ended with fists flying outside in the alley. In short, the show usually presents the sort of quasi-sociological dabbling that will warm the heart and titillate the emotions of any socially conscious viewer without really affecting his mind or his deeper convictions.

On "East Side/West Side" the continuing action evolves around a social work center, and the principals are a curious lot. The center is supervised by Elizabeth Wilson, who deports herself like an aging debutante doing penance for inherited wealth by performing humanitarian tasks she secretly considers beneath her. She is helped by Cicely Tyson, who is a good actress but who is handed a part so shallow as to confirm everything derogatory one has ever heard Negroes say about Negro social workers. But the central character is acted by George C. Scott, and it is he who gets most of the attention.

Scott is an exceptionally well-equipped realistic actor: he sits in his office chair or he crosses to a filing cabinet in such a way as to let us know it is an action he has performed hundreds of times. If he is preoccupied, he is off-handed with his lines—as if he were preoccupied.

On the other hand, Scott is clearly star material, although he may never be anybody's matinee idol, and the attention that he gets therefore has more to do with the personality he projects than with his talent. Much as another promising realistic actor, Marlon Brando, projected something of a petulant baby, Scott projects a petulant adolescent. The analogy may go even further, for unless Scott is able to avoid the paraphernalia of stardom, as Brando was not, he may find himself in the same artistically frayed position that Brando is in now. And personally Scott, like Brando, seems to have strong notions about what is and what isn't worthy of his time and talent.

If Scott does find his way out of the trap of fame and stardom, if he does find a way to carry his talent with effective dignity, he will probably have done so for himself, for certainly there is little in our culture and little in our theatre to help him. Meanwhile, if "East Side/West Side" can observe the city and its people, in the honest dramatic journalism of "Who Do You Kill?", letting the art fall where it may, it will have a job worth doing.

(Village Voice, January 9, 1964)

A Boone to TV?

The "Richard Boone Show" is worth discussion for what it promised to do. Boone announced he would be presenting repertory theater on TV, with a sustaining group of players, with an emphasis on good scripts, and with the late Clifford Odets as story editor. The resultant showy façade hid a self-deluded series which seems worth discussion as an example of what fatuous or highfalutin' or even guilt-ridden ideas many educated Americans cherish about something called Art.

On the Boone series, repertory theater meant using a continuing cast of character actors of various ages, sizes, shapes, ranges, and degrees of talent. These actors changed costumes, make-up, and (if the player could manage it) accents from week to week. (Those make-up jobs, one should note, were sometimes so rudimentary that they might embarrass a Senior Class Mummers Club.) The emphasis, then, was on the surface paraphernalia of characterization, and acting easily became gimmickry.

And the scripts. On the opening show a bully of a DA, played by Boone, got the truth out of a wife, played by Bethel Leslie, that she had shot her husband. The DA discovered, in her relationship to the husband, a supposedly crushing parallel to his own marital situation—all of which was not only mechanical but duly telegraphed early in the script. Boone was a convincing bully, as usual. And he was unconvincing as anything else. Miss Leslie handled her *tour de force* with aplomb, but a couple of difficult character transitions were undramatized, being simply assumed after station and commercial breaks.

In "All the Comforts of Home," a man from the Southern mountains returned home after years of trying to make it as a city hillbilly singer, disillusioned and careerless. At the end of an hour this man has a wife (a woman who is held suspect by the rest of the community), a blind step-daughter, and he is disillusioned and careerless. Actually, the writer had hold of the wife's story, but he didn't seem to know it. In "The Stranger," a clumsy fantasy by Dale Wasserman, we watched a group of decadent people confronted with a totally innocent young man, each trying to mold the youngster according to his own failings. We were left with the conclusion that one lives his life in childish innocence or in certain corruption. And in "Wall to Wall War" we saw a berserk, machine gun-wielding insurance salesman turn his office into a battlefield, either by psychotic delusion or psychopathic cunning (the author didn't seem to know which), and turn his fellow-employees into an attacking platoon, wounding some of them in the process. This stridently crude playlet had a two-minute opening and a five-minute ending, a declared lesson about what society had done to this fellow, several years before, in having trained him in cruelty to help fight a war. In between it was admittedly interesting to watch the actors unwinding one absurd bale of padding after another. In "The Fling," an invalid wife forgave her otherwise devoted husband a fall from fidelity by changing the subject as he is about to confess. (*Redbook* anyone?) The nadir was "Where Did You Hide an Egg?" a sort of slapstick *Treasure of Sierra Madre,* handled so badly by Boone, Harry Morgan, and Robert Blake as to make one feel that the Three Stooges are superb comic performers by comparison.

The series had one good show, "Don't Call Me Dirty Names" by John Hasse, in which the author fleetingly showed the way out of the maze of fatuousness; you have to accept popular dramatic conventions and make something out of them. The playlet began as a chase thriller, with a father trying to locate his daughter who is pregnant, unwed, and apparently bent on suicide. He finds her about halfway through the script, whereupon Hasse managed to sustain and extend his suspense from chase into psychology as the father tries to help the girl pull herself out of despair. Hasse began with a conventional popular form, even using the currently fashionable ingredient of the unwed mother; he wrote for a mass medium and for an arty anthology series within that medium. All these things might have tempted him to moral, psychological, or dramatic fudging, but he fudged not; he may have been conventional but he wasn't fake. As the father, actor Lloyd Bochner was handed an extremely difficult part. One false note of self-righteousness, sentimentality, or hardness would have destroyed Hasse's effect. Bochner struck not one.

No doubt admirers of Clifford Odets will assure themselves that he died before approving much of the gimmickry and moral posturing which this series offered. But since Odets is not my idea of a perceptive or even dramaturgically sound playwright, it would not surprise me to learn that he had approved most of these scripts. Odets contributed two playlets himself. One, "Big Mitch," a domestic tale of an aging man living out of some self-important delusions, might have been done by any of the several Odets-Saroyan imitators who used to monopolize the Studio One, Kraft, and Philco theaters. The second, "Mafia Man," was an anonymously written thriller.

With such an approach to acting, with a grab-bag of scripts, and with a frequent shift of directors, the series offered, not repertory theatre, but stock company charades.

As a final irony, several of Boone's early Paladin shows—those which made an honest attempt to discover the meaning of a threat of force in maintaining social order and public morality, and those which tried to relate intelligence and violence—would surpass most of what we have seen on the Boone theater.

Art, it should go without saying, is not something you produce by intention or by taking a stance. And it is certainly not something you undertake in contrition after you have made your loot doing a couple of TV series you apparently don't approve of.

(*Kulchur*, Winter 1964-65)

Gore on the Living Room Rug

I recently wrote to the editor of one of our better-known middlebrow culture magazines proposing an essay in defense of television. What I had in mind, I said, was the unique style and often fine comedy of certain cartoon shows ("Bullwinkle," "Huckleberry Hound," "Quick Draw McGraw," etc.), the strikingly different moral outlook of certain TV Westerns (was there ever a man like Matt Dillon Saturday mornings at the Bijou?) and the terse dramatic energy often achieved on "Naked City."

The editor's reply was swift and blunt. Television, he asserted, matured quite early with "brilliant" writers like Paddy Chayefsky and Tad Mosel, but it is already in its decadence, as witness its constant resorting to crime, violence, and sex. He knows because he has to watch a lot of it as a member of a committee which offers annual awards for good television.

I could not resist an immediate reply, and I wrote to him more or less as follows:

Dear Sir: Thank you for your prompt answer to my letter proposing an essay in defense of television.

Although I can't quite agree with you that the view of TV I suggested was "rosy," it does not surprise me that your feelings about the medium are not the same as mine. However, I am allowing myself a certain disappointment that, to judge from the tone of your note, you will brook no disagreement. I might add that there are those of us who feel that Chayefsky and Mosel do not really belong on TV but have found their metier on the Broadway stage, and who found the recent retrospective of early TV shows, given at the Museum of Modern Art, a distinct embarrassment.

On your remarks on the violence on television, let me tell you briefly about one play I watched. It opened with the hero having hallucinations and brooding about killing (there had already been one murder). Soon this same young man was exhibiting an incestuous yen for his mother in a bedroom scene. He also bullishly insulted a well-intentioned elderly man, and later stabbed him and dragged his body around by the ankle. Then he abused and slapped his girl-friend until she had a mental breakdown and finally died. At her burial he morbidly fondled some bones somebody had dug up in the cemetery; then he jumped inside her grave and tried to make love to her corpse until they had to drag him out. The thing ended in a violent scene involving the hero in stabbings and poisonings until the TV screen was littered with dead bodies, including his own. At this point, a secondary character turned to the hero's corpse and said something about "Good night sweet prince!"

Well, I tell you it's no wonder our young people are so shook up these days, with this sort of thing coming into the home. (Sincerely yours)

My letter may have made me feel better, but I doubt if it had its desired effect. "Do you really mean," I could hear him saying, "to compare 'Have Gun, Will Travel' and *Hamlet*?"

Well, yes I do—or I might mean to compare such early Elizabethan tragedies as *Gorboduc* or *Cambises* and "Have Gun." And I might compare many a TV comedy show and early comedies like *Gammer Gurton's Needle* or *Ralph Roister Doister,* not to mention many a *commedia dell' arte* scenario.

Hamlet could not exist without crude and violent theatrical vengeance and crude buffoonery in its background. Nor does *Hamlet* exist aside from its own violence. It exists in coming to terms morally with its violence—which is to say with human violence. Gloriously, it suppresses nothing.

I am quite convinced that if we could present our editor with *Hamlet* itself, but somehow in the guise of a contemporary television play, he would be shocked and horrified. And so probably would any Elizabethan follower of either Sir Philip Sidney's dramatic criticism on the one hand or the Puritans' attacks on the stage on the other, be outraged by *Hamlet*.

Our editor's final point would undoubtedly be that television has a duty to put on performances of Shakespeare. The more good performances of Shakespeare, the better to be sure. But television's primary duty, like the primary duty of the Elizabethan stage, is to itself as a unique dramatic medium. It must find its own way, not only against our middle-brow reviewers and our Puritans but also by stumbling and clawing across Hollywood production associates, network vice presidents, and ad agency men, and get at the truth. And in getting at the truth it must learn to come to terms with violence and sex.

I am not saying that TV has produced any *Hamlets* but I believe it has already produced its share of dramas comparable to *Hamlet's* best precursors, its share of *Mucedoruses* and even *Spanish Tragedies.* *

(*Kulchur*, Winter 1964-65)

Proud Gleanings

"Espionage" was publicly lost in the ratings piled up by "The Beverly Hillbillies," by "Ben Casey," and (those often deserved) by Dick Van Dyke. Unfortunately, I can't discuss the series without at least mentioning a shocking early segment called "The Whistling Shrimp." It began as a stirring indictment of what the CIA might become (or quite possibly is) as its activities determine policy. However, the playlet ended in a cowardly four-minute retreat: "Sure, sure, we arranged to have the elected leaders of the country killed but, after all, we saved that country from Communism!"

Otherwise, "Espionage" presented one really commendable playlet, one of the best ever seen on television. It was "Medal for a Turned Coat," written by Larry Cohen and movingly acted by Fritz Weaver. It had to do with a German officer who had gone to England in 1944, proposing a truce be arranged between the anti-Hitler faction in the German Army and the Allies, and had remained there. As the action began, this man was returning to receive a medal

---

* This is as good a place as any to remark on the revival by the Public Broadcasting Service in 1980 of the original *Requiem for a Heavyweight*, an event which surely represented a full circle of some kind. One could not fault the acting or staging or direction on that kinescope surviving from the 1950s. But the script and its characters are indicative of the kind 1950s outlook of New York TV, and they reflected the taste of its critics as well. Rod Serling's boxer-protagonist is evidently inspired by Arthur Miller's Willie Loman in *Death of a Salesman*, and, like him, is something of a self-deluded pathetic. For him one can feel sympathy and condescension all at once. He is a pseudo-tragic loser who stirs within us a warm glow of patronizing self-satisfaction. As James Agee said of Miller's *All My Sons*: Ibsen for the self-righteous.

from the West German government. Through a free, flashback, interplay of past and present time we learn that he had written a book exposing "the Nazi character." We see him confronting his sister who knew him when, and who has read his book; confronting an adolescent nephew who knew nothing of wartime Germany; and confronting the nephew's friend who is a brooding neo-Nazi. We see him meet the Prussian general who had originally sent him to England. "Would you really have defected," he has to ask himself finally, "if you were not convinced that Hitler would lose?" Most of all, we see him crucially confronting himself. I should not say more about this exceptional script, since Cohen has so tellingly integrated his action with his meaning.

Cohen borrowed some of the quasi-cinematic devices of *Death of a Salesman*, but his hero is more complex and more capable of self-examination and moral insight than Miller's; his secondary characters are also more complex; and his action, it seems to me, more universal in its implications. "Medal for a Turned Coat," in short, came very close to true tragedy.

Another "Espionage" installment called "A Free Agent" featured Anthony Quayle, surely one of the most remarkable actors alive.

A "Defenders" show called "Moment of Truth" was also eminently worth seeing for the exceptional and keenly observed character performances of Jack Guilford, Anne Jackson, Norman Fell, and David Karp.

An "Arrest and Trial" show, "The Revenge of the Worm," had an exceptional performance by an exceptional actor of decidedly exceptional range, Telly Savalas, who can be convincing as anything from a Near Eastern prince to a Brooklyn trucker. It also had a startling scene in which an unbilled actor, playing a gangster, combined humor, stupidity, and cold menace with simultaneous subtlety. The script of this show was an interesting effort to deal with the consequences of "justified" personal violence against an admitted evil. It was hampered by what I judge to be the tampering vacillations and machinations of a Hollywood story editor, but the writer's main points were not altogether obscured.

The Alfred Hitchcock shows during the 1963-64 season were so bad that they were often incomprehensibly ridiculous. However, "Beast in View" was a good, slick job, for all its transparencies in plotting, and it featured another commendable performance by that remarkably underrated actress, Joan Hackett.

These being some of the distinct pleasures I had watching television in 1963-64, I don't feel disposed to demand much more of the medium. Has Broadway or the movies offered as much?*

(*Kulchur*, Winter 1964-65)

In Memoriam

Summer TV, we are frequently told, is a bad time in which endless re-runs vie with cheapjack summer shows, featuring minor talent. It is—or rather it was. It was also a good time for catching up and for looking into unlikely corners. In so doing, I ran across some interesting, worthwhile shows. They will not be seen again—many of them—in network runs, but most of them will rerun on local channels. In any case, they deserve recognition, and as far as I know, they have received little or none.

"Lancelot Link/Secret Chimp": This kiddy show is a frequently hilarious spy spoof, acted out by a group of well-trained and carefully

---

* One of the privileges of writing regularly about TV in mid-1965 (had I had it) would surely have been writing about "The Man from U.N.C.L.E.," its cool and frayed heroes, its earnest and innocent heroines, and its plots, which fun-loving writers could manipulate in almost any direction by having a few well-chosen words spoken into a ballpoint pen ("Open channel D"). And on the British import "The Avengers." In the midst of one typically sinister, convoluted plot, heroine Diana Rigg did a complex, piquant exposition scene, while lounging around Steed's flat, and intermittently trying out some notes on a tuba, which just happened to be there.

photographed chimpanzees. Dull, bungling Lance is assisted by Marta Hairy, who has to be seen slinking around in bow-legged mesh panty hose (and speaking with the dubbed-in voice of a faulty brake-lining) to be believed. If the plot gets bogged down, there's the Evolution Revolution, a burlesque, simian rock group, complete with long hair, battle jackets, and a gleeful drummer bashing away to the rear.

"The Senator": This one *was* praised of course. It ran as an occasional segment on "The Bold Ones," got an Emmy award, and has been cancelled for next season. There were few bones made about the liberal viewpoint of the leading character, but there was occasionally a clear exposition of the intricacies of American politics and the claims of politicians, local and national. And the show takes a frequently unflinching look at the moral issues raised in its stories. Casting, performances and photography are executed with a rare vividness and immediacy, and the shows' producers and directors often turn the usual TV production corner-cutting camera work into expressive techniques.

"Green Acres": Another cancellation, but after several seasons, and one definitely set for local channel replay. Don't identify this comedy about a New York couple transplanted to the farm with the simple rusticity of "The Beverly Hillbillies" or the sticky, self-satisfied cuteness of "Petticoat Junction." This show joins farce and whimsy (a difficult trick). Eva Gabor's version of a middle-European Gracie Allen either works well or it approaches the extremes of cyclamate (and you know what happened to that). Eddie Albert is superb. If he showed one more degree of exasperation and anger at his bungling but triumphant bumpkin antagonists, he would become a bully or a bore. But he never does. (Those antagonists, by the way, include a talking pig named Arnold whom everybody understands but Albert.)

"Can You Top This": A syndicated show that runs on local channels only. If you enjoy watching a group of pros delivering, embellishing, and generally horsing around with some jokes that you may or may not have heard before, then you will enjoy this casual, highly informal half hour.

"Here Comes the Grump": Another kids' show, this one animated and especially good for those who enjoy the kind of adventurous whimsy that is featured in the later Oz books.

(*Herald,* September 19, 1971)

Invasion of the Hollywoodites

Whatever else may be said about the network TV season, 1971-72, one thing is certain: there will be less of it. The FCC has decreed that local channels must carry fewer network hours, and as a result, NBC, CBS, and ABC evening programming will begin a half-hour later, except on Tuesday for some reason.

The FCC action is the result of a bit of wholly misguided idealism. The agency wants to encourage more creative, locally oriented programming. Nobody else, absolutely nobody else, believes it will. Local channels aren't producing their own shows. They are buying up inexpensive taped shows, produced for station-by-station syndication; syndicated outfits have been working overtime during the summer to grind out inexpensive quiz shows, stunt shows, and continuations of network rejects ("Hee-Haw!" "Lawrence Welk," and "Lassie" are all having second lives via syndication). And small outfits are dusting off their ancient TV films for local replay. (If this keeps up, "My Little Margie" may rule nighttime TV once again.)

Otherwise, this TV season will be known as the time of the movie stars—or of those established movie stars who apparently can no longer make it at the theatrical box offices and so are trying it on TV.

James Stewart, cozily renamed Jimmy Stewart, is a college professor, a father and a grandfather (he has a son the same age as his grandson, eight). Rock Hudson is a carefree, married detective (if that's not a contradiction *à trois*). Shirley MacLaine is a photojournalist. Anthony Quinn is a Latino mayor in the Southwest. Tony Curtis is a non-smoking millionaire who, with Roger Moore, sticks his nose into international crime. Glenn Ford is a contemporary Western marshal who saddles up a Jeep. And George Kennedy, who, it is to be hoped, will quickly sell his Hormel and Kingham stocks, is a priest who solves murders (consider the possibilities).

Then there are the TV returnees, players who have had a fling at films and by now presumably worn down their box-office potential. These include Dick Van Dyke, David Janssen, Rod Taylor (one of the most convincingly energetic actors ever to work on TV), and James Garner.

How all this will turn out should surely remain to be seen. One of the most pointless and foolish tasks in TV reviewing seems to me the business of pouncing on new shows after one or two episodes have aired and delivering pronouncements thereby on a whole series. I don't like to remember what I said about the earlier Dick Van Dyke series after one show, or even after ten. Yet he went on to develop a fine comic acting style, perfectly suited to the TV medium. (Anyone who thinks Jack Lemmon is one of our best performers of comic drama might study Van Dyke carefully, and think again.)

Meanwhile, there are the returning shows. And since so many of the new ones are suspense and/or mystery shows, it might be well to say something about a couple of such programs which I may have mentioned here before, and which seem to me to set standards in that tenacious American genre.

There's "Ironside" with Raymond Burr. Basically, this one offers unusually well-produced and well-acted plot-mongering suspense melodrama in which story per se takes precedence. When it rises above this, it often does so because of Burr himself and a script that lets him show what he can convey about the conflict of human feeling and human reason in a man of superior insight and sensitivity.

Then there's "The FBI." The bickering about whether this one is or is not a propaganda work for J. Edgar Hoover and his forces seems to me largely beside the point. It is a good show because its wrongdoers are not one-dimensional badguys, but are often presented as beings with strong, believable personal conflicts, temptations, and moral choices to make, and because its situations sometimes bring its characters through the fire of self-knowledge—the father of a kidnaped son must decide what he *really* thinks of his adored boy; a young radical is forced to ask himself just how much moral conviction is really involved in his political stand.

What all this has to do with the real FBI and its actions, I don't know, and I don't think I care. Indeed, the agents themselves are apt to be plodding, one-dimensional goodguys. But that tendency is often saved by the performances of Efrem Zimbalist, Jr., whose understated style makes him an exceptional TV actor in ways I hope to discuss soon.

(*National Review,* October 22, 1971)

Open Letter

Dear Laura Z. Hobson:

I have read your meticulously researched article in the *New York Times* critical of "All in the Family," and I must say that I find myself sympathetic to your feeling that the show pulls its punches and is indeed in danger of making a lovable fellow out of its chief bigot, Archie Bunker. That position has been taken by others, as you may know, particularly in England where "All in the Family" had its origin in a series called "Till Death Do Us Part," which was apparently stronger stuff.

I am told also that graffiti promoting "Archie Bunker for President" have appeared in the men's rooms of certain workingmen's bars in New York City. Food for thought, even though a perhaps more serious "George Wallace for President" may be scrawled in the next booth.

However, the tone of your article makes me wonder how you respond to comedy in general, and to good TV comedy in particular. How do you like the "Mary Tyler Moore Show"? What do you think of "Lucy"?

More to the point, what is your response to the kind of Jewish jokes that Jews tell other Jews? and the jokes that blacks tell other blacks? What do you think of the comic nature and function of such humor? More important perhaps does it make you laugh?

I appreciate your quotation from Henri Bergson on laughter—one should of course treat them with great respect. But is it not interesting that we have such general agreement on the nature of tragedy—an almost unbroken critical concurrence that begins with Aristotle—but have had so little agreement among theorists on the nature of comedy and humor and the function of laughter? However, there *is* some agreement on Aristotle's precept that comedy shows men as less than they are.

Take a couple of extreme subjects: We have had good comedies about prostitution and good comedies about murder. But a good comedy about prostitution must not show the extremes of lust and of degradation that each of us knows is a part of the world's oldest profession. And a comedy of murder must not show us blood and gore on a murderer's hands.

Note also that a great comedian, Charlie Chaplin, even found Hitler a suitable object for comic ridicule. Is it appropriate to add that Chaplin is Jewish?

I wonder if it is possible for "All in the Family" to ridicule Archie Bunker's bigotry to an American audience in 1971, if Archie were (as you suggest) to speak of "kikes" and "niggers" as real bigots do. Or is it perhaps that you really don't want a comic treatment of bigotry? Is bigotry too sacred for ridicule?

(*National Review,* November 19, 1971)

PBS at Random

All I know about Public Broadcasting is what I see on New York's Channel 13. That, and what I read in the papers. I get a lot of the latter (or I could if I wanted to) because educational television seems to be almost the only kind that the fellows at the *New York Times* find worth writing about. I don't get much of the former because I don't watch much Public TV. Not because I don't try to. So the remarks that follow will be limited, tentative, and, if you will, biased. I'm not even sure of the difference between old NET and new PBS, except that the former had foundation support, the latter gets government funds, and neither of them is necessarily Educational, or Public.

Take "Hollywood Television Theater," a series of highly praised "serious" dramas that most of the good actors I know of are either dying to get on or proud to have been on. I tried to watch Eli Wallach and Anne Jackson do *The Typist* and saw two capable players, stiffly directed, and stridently projecting all the way to a top balcony that wasn't there. (Couldn't they *see* those TV cameras a few feet away? Don't they *know* that people keep their TV sets just across the living room?) I switched away to a network whodunit, and although they weren't doing much, they weren't pretending to, and they were doing it expertly.

No, I didn't watch the "Forsyte Saga," or any of its successors from the BBC, and that puts me in a tiny minority among my acquaintances. It's interesting that all sorts of people who say they never watch TV, and certainly never watch soap operas, watched *that* soap opera. Is there a parallel here to the thousands of American parents who don't allow their kids to read comic books but who are delighted to have them read "Tin Tin," a Belgian adventure strip, if it's translated in *Children's Digest* by *Parents Magazine?*

More to the point, it is interesting that some of the most effective and praiseworthy TV that PBS has given us—"Sesame Street," "The Great American Dream Machine"—has learned important lessons of presentation, format, pacing, and tone from commercial TV, and learned them well.

To be sure, I'm grateful to Channel 13 for broadcasting some of Frederick Wiseman's alarmingly revealing documentaries ("High School," "Law and Order"). And for whatever music PBS broadcasts (although music on TV presents very special problems that have rarely been confronted and never been solved). I admire the way Julia Child's voice jumps that fine interval (about a third, isn't it?) with such natural ease. And I have no particular objection to the way Robert Cromie bulls and charms his way through all those half-hours with Famous Authors. In general I suppose I'm glad our Channel 13 is there, although I find its genteel, middlebrow cultural supermarket frequently deplorable, not to mention uninformative.

Lately we have been offered amiable junk, *Son of the Sheik;* a comic masterpiece, Keaton's *The General;* and a creditable diversion, *The Extra Girl,* all laid end to end as Motion Picture Art—and sometimes run arbitrarily at "sound speed" (24 frames per second) which can turn a superb and significant spectacle like *Intolerance* into a jerky incomprehensibility.

At this rate, years hence PBS will be scurrying around to find prints of Television Art, and (to confine myself to recent shows) we can hope come up with the time Carol Burnett did a marvelous Busby Berkeley spoof, "43rd Street," or the night Mary Tyler Moore's boss had to take over on-camera news duties during a strike, or Dick Van Dyke took up motorcycling and found himself (he thought) dying in the desert, or the recent "Gunsmoke" on which well-intentioned whites nearly kept an ancient Indian from dying in dignity, or find anything in which actors like William Windom or Geraldine Brooks or Lee Grant or Charles Aidman build strong, sometimes touching, characterizations out of little more than the devices of a melodramatic plot.

Personally I don't want to wait for such rediscoveries. I'll take my chances on Channels 2, 4, 7, et al., and watch such small, unembarrassed examples of TV art right now.

(*National Review,* March 3, 1972)

Flipping In

Flip Wilson may be the most intelligent clown television has ever seen. (I am tempted to say that he is the most intelligent clown that has ever been, although obviously I can't know that.)

His intelligence is something that we discern of him behind his clown's mask, of course, and, accurately or not, we always discern something behind that mask, for clowning is a thing of double images. We are not at all surprised to learn that Stan Laurel, who played the simpleton on screen, was an industrious craftsman off screen, who wrote and directed films besides clowning in them. We would guess that Red Skelton is a sentimental and possibly maudlin man off camera. Or that clown C is an egomaniac. Or clown D an arrogant and rather humorless man with his mask off. But Flip Wilson seems so intelligent, so dignified, has so much self-knowledge, that it is a wonder he can clown at all. Yet he clowns uninhibitedly and excellently.

That is not all he does. Flip Wilson can do an emcee job in his own person. He can do a stand-up comic's string of quick gags. He can do a storyteller's routine of rambling, embellished comic tales. He can banter with a man like Bill Cosby, as he did on his first TV season, semi-improvisationally and with both of them showing an impeccable sense of comic timing—the kind that old-timers always claim takes decades to develop. And he can play straight man to somebody else's clown.

But best of all, he can clown himself, whooping and prancing as the Reverend Leroy, a fine combination of revivalist preacher and

con-man. ("I know him very well! We got the same parole officer!") As Freddy Johnson, the fumbling would-be playboy. As a
sentimental short-order frycook. And as Geraldine Jones.

Geraldine is a wonder. How can a man in drag, no matter how
funny he is, help but be embarrassing? Flip as Geraldine isn't one
bit embarrassing. And Geraldine's combination of apparent dumb-
ness, shrewdness, independence, and total, spontaneous self-accep-
tance—and blackness—make for one of the most complex comic
masks we have ever had. The classic line "the Devil made me buy
this dress!" is hilarious, not only because (like all good comic lines)
it is at once both unexpected and entirely fitting, but also because it
says so much more, and so much less, than it seems to say: it explains nothing, excuses nothing, promises to reform nothing. It is a
kind of benign hedonist boast.

There was one Geraldine sketch the first season which, using
some old and some new material, was such a total comic put-down
of all contemporary attitudes that it was hard to believe that it—or
any part of it—was on TV in the 1970s. Geraldine had taken a job
in a rather lavender hairdressing salon. In comes Moms Mably,
querulous, irascible, but interested—her false teeth out, clouding her
speech, as usual. Naturally she gets stuck in the back room with
Geraldine. But it turns out she is a maid (!) to a wealthy matron
who wants her "done over," and at that point the ambitious, effemi-
nate, white proprietor tries to take her over.

Geraldine—hand on hips, head asway, eyes afire—lets him have it:
"Don't you *touch* my customer! Don't you *ever* touch *my* customer!
Look out!" *Et cetera,* ending on a muttered "You and your equal
opportunity employment." Exit proprietor. But not before the contrast between Flip's naturalness as Geraldine and the "fancy hairdresser" affectations of the owner have become both blatant and
intricately complex.

Nothing pleases Moms. Finally Geraldine crosses to the wig-
stand. "Honey, you gonna love this. This is called the Afro-natural."
She adjusts the wig on Moms.

And Moms eyes herself this way and that in a hand mirror before she asks in a bass rumble, "Sweetheart, do a spear go with this thing?"

Which of us, in the 1970s, can quite recover from that?

                                                          (Unpublished, 1972)

III

Elaborations

Ike, Mr. Dillon, and the PTA

President Eisenhower is supposed to have admitted recently that Sunday afternoon television bores him, and that means that Ike and I feel the same way about at least one thing. By any meaning of the word I know the *cultural* desert of television comes on Sunday morning and early afternoon when the culture shows are broadcast and we are treated to the culture of a middlebrow American intelligentsia. A cracked-voiced Midwestern college professor, for whom learning and charm are apparently synonyms, corrects our speech under the premise that whatever is is right so long as one can be erudite about it; a balding English college professor turns the sometimes pointed comments of his sometimes interesting guests (experts on everything from Laurel and Hardy to beatnik perversions of Zen) into a comforting blandness by his oh-so-interested responses to them; and later in the day two heavily-eyebrowed network newsmen can be counted on to parlay some old newsreel clips into the most thrilling melodrama in town.

Sunday afternoon cultural television already has the sanction of *Harper's* magazine, and it can count on being congratulated regularly by Jack Gould and John Crosby. Perhaps all it needs is David Susskind to come in and uplift its intellectual level with some of those flabby pastiches of old movies he rewards us with every so often in the name of high drama—that is, unless what it really needs is Susskind himself to lead a discussion. He's even better than our professor; he can make intellect, artistic dedication, and the problems of world politics come out as just good old smiling charm. At any rate, you can be sure the whole thing will please the National PTA which has declared itself opposed to all violence, lust, etc. (you know the words) on television. I suppose Maurice Evans and Hallmark greeting cards had better get themselves another gimmick because the PTA obviously stands firmly against Shakespeare. And "Lamp Unto My Feet" maybe better cool it on the Bible for a while too till this thing blows over.

There is a remarkable time on television, however, and I recommend you watch it. Indeed, I doubt if there has ever been anything quite like this hour-and-a-half in mass entertainment (or in American *culture* for you Eisenhower fans). Not in over fifty years of the movies and not on the radio. It comes on Saturday night, and it is three "Westerns."

I put the word in quotes, since the real point of these shows has little to do with ridin', shootin', rustlin', ropin', or Dead Gulch. The West is there, really, because it is one of the few areas in mass American fiction where a moral issue is raised honestly and where a tragic view of the human soul can thrive. The other is big city gangsterdom. If the West of "Gunsmoke," "Have Gun, Will Travel," and "Wanted: Dead or Alive" has nothing whatever to do with the "real" West as it existed, no matter, for it can have something to do with each of us and I think that does matter.

That hour-and-a-half, those three shows constitute, as I say, one of the most remarkable achievements in mass-produced art we have ever witnessed. Basically what the programs offer are vignettes of character, clashes of personality. When the clash is based on the merely eccentric (and it often is) the result is inevitably trite and trivial. When it is not, there is a remarkably honest effort to face the moral issues the script raises and see them through; the writers in their sometimes sardonic way seem compassionately committed to their work and their characters. The old class "A" film script might raise such issues but would seldom be willing to face them as these scripts do. A radio script would seldom raise them. A class "B" film script might raise comparable questions and even see them through but only in terms of a melodramatic plot with such all-good and all-bad characters that the point involved is too obvious for any moral doubt to exist. I suppose the highest tribute one could pay one of these shows is to find himself asking, "Just what *is* the right and wrong here?"—and he may ask himself just that during a "Gunsmoke," a "Have Gun," or a "Wanted" episode. The appearance of a Western marshal-hero who has to make moral decisions, who can make wrong ones, and who knows it in a TV series is certainly a cultural event of some significance. A gunfighter who is, for all his sometime smugness, capable of seeing, accepting, and sometimes carrying on his own shoulders the sins of his lesser fellow men is, potentially, an epic or tragic figure of magnitude. The weakest hero is the bounty hunter on "Wanted." There is an implicit cynicism in his character that is more difficult to balance than that in a Faustian gunslinger, and (listen to this, PTA) there is often a consequent violence for its own sake, sadistic, extrinsic, unpurged, unresolved. Such is the natural result of an effort to compensate for the hero's job by a sort of "maybe he's making a living off bounty hunting, but look at the people around him—they're worse." What might really redeem such a man is an implicit depth and self-doubt and seldom do the scripts or the lead actor (Steve McQueen) quite provide it.

The scripts don't preach; anything they have to say springs out of their action and conflict. Some of the things they seemed to me to be saying recently were: being agreeable to people is not necessarily the highest moral call we may get—in fact, being agreeable may have its evil side. Knowing what is best for someone else is frequently vicious. A man who is strongly judgmental of the motives of others had better look to his own. Love is giving, not getting.

Truisms? Perhaps, but truisms often unlived; lately unspoken; and in our entertainments, usually contradicted by the wish-fulfillments. To think that these episodes are produced at the rate of one a week, thirty-six a year; almost astonishing!

So, Mr. Susskind, please don't tell me again that TV Westerns have the same old predictable stories, because I happen to know that a predictable story is a good working ingredient of tragedy: everybody knows "what happens" in tragedy, everybody is supposed to, and it has always been that way with tragedy. The dramatic question is *how?* and *why?* Your question of *what?* is the melodramatic question. And if I really want something predictable, I can always catch those cultural discussions on your "Open End."

I want to be saved as much as the next man, but I don't think the escapist entertainment of English etymology or a group of Broadway intelligentsia sitting around gassing about "too much violence on television" over coffee is going to do it. I like the old way where something makes us realize that there is a lot of violence (or meanness or blindness) in each of us, so somehow we have to learn to deal with it with a little compassion for everyone involved.

The Westerner has been the most enduring stock character in American fiction for over sixty years for one reason: his being carries the most complex meaning for us all. We may hope that someday a major writer is going to realize it. When he does, he will be thankful for the important work of clarification, focus, and refinement that his predecessors in American *culture* have done for him.

(*Kulchur*, Spring 1960)

Clownish TV:
I—Dick Van Dyke

In the early 1960s, Channel 13 in the New York area ceased to be a northern New Jersey commercial outlet and became New York's educational television station, and one of its first shows offered some very early Chaplin two-reel comedies. When these innocent little knock-about farces were first made, it would surely not have occurred to the members of their delighted audiences that one day they would be considered "educational." It probably occurs to no one today that the best comedy shows from commercial television will some day also be considered educational, but if there is such a thing as aesthetic justice, they surely will be. And in perhaps twenty years, educational television will be presenting the best of Lucille Ball, the best of Jackie Gleason, the best of Phil Silvers's Sergeant Bilko, the best of Milton Berle. And, again if there is such a thing as aesthetic justice, educational TV can turn to the film library of the Museum of Modern Art for prints of these shows.

I have held back mention of one outstanding TV comedy series, the "Dick Van Dyke Show," and the shows of one comedian, a man who has lasted longer than any other on TV, Red Skelton. Each series is worth discussing not only because it has produced good shows but also because it has managed to reveal something of the nature and possibilities of the TV medium. These shows and the others on my list are television. They are not warmed-over Broadway. They are not re-run movies. Nor are their techniques simply borrowed cinema.

The Van Dyke series is an example of what the trade calls "situation comedy," and the history of that genre is interesting. In the early days of radio, most popular comedians told jokes, held conversations with second bananas, singers, and band-leaders-turned-third-bananas, and did a couple of brief sketches. The idea of a situation comedy developed when it became obvious that a show which used a constant barrage of snappy jokes week after week would soon wear itself out. Performers, writers, and producers hit on the idea of using an extended situation, a miniature comic plot sprinkled with jokes, with the same characters from week to week, and lasting half an hour. Instead of keeping the audience interested merely by a flow of cracks, the comedian could coast on the comic plot as well, and still come up with an entertaining show.

Inevitably, a kind of "family comedy" appeared as one format for a situation series, and, once that sub-genre was set up, a core of conventional plots and situations grew out of it. I realize a reference to *commedia dell' arte* can be death to discussions of theater and theatrical conventions. But the parallels between its practices and the conventions that grew up in radio and TV situation comedy are inescapable. I say parallels rather than influences with some care, not only because one could not prove the influence of *commedia* on TV comedy, but also because one does not need to. Its stock characters and set situations are almost exactly those of Japanese Kabuki, East Indian comedy, and the Astaire and Rogers musicals of the 1930s. Which is to say that they are comedic and human universals.

It is impossible for an outsider to know what the Van Dyke show was like when it was first conceived by writer-comedian Carl Reiner as a vehicle for himself. What one can say is that as it came out with Van Dyke in the lead, and as a Sheldon Leonard–Danny Thomas production, it was another link in a chain of comedy sausages that also held Thomas's own show, Andy Griffith's show, Joey Bishop's show, and (with a slight switch here and there) a couple of others. But the Van Dyke series was link sausage only as it first appeared and only in its basic set-up. That is, it had a husband and wife, a child (under ten, precocious, but not too), and one or two lower comedy second bananas.

Like the Thomas show, the Van Dyke series had a dual home life–showbiz background. Thomas played a comedian with a family. Van Dyke, as Rob Petrie, worked as head of a team of writers for a successful TV comic. (I probably should not speak of the series in the past tense since it is still re-running.)

I note that one highbrow commentator has dismissed the Van Dyke show on the grounds that Mary Tyler Moore, as Van Dyke's wife, Laura Petrie, is prettier, more quick-witted, and more vivacious than any American housewife could possibly be. Tsk, tsk. One wonders what the gentleman would make of Rosalind or Beatrice, not to mention Cleopatra. One should add that Laura Petrie, as Miss Moore played her, was sometimes shy, awkward, embarrassed, and despite the best will in the world, comically and totally incapable of hiding her true feelings about anything. Another of our highbrow critics of television declared that the show reflects middle-class values. Indeed it does. And, one might add, how much great comedy does also. Rob Petrie, as Van Dyke played him, was something of an intellectual but sometimes shy, sometimes cowardly, and often afraid of his boss and afraid for his job. But *must* one invoke Aristotle to answer criticism and maintain that a comic hero is by nature smaller than life?

My own reaction to the show's half season was that the formula was being given too conventional and sentimental a ride. And I felt that such redeeming features as Van Dyke's very good, partly pantomimed routines—as a man who comes home sloppy, falling-down drunk but who succeeds in hiding it from his strait-laced wife by a series of quick, hairbreadth recoveries to a stiff semblance of sobriety every time she re-enters the room; or as a man giving one of those earnest "public service" talks on safety-in-the home while managing to slam or prick his fingers with every drawer, paperweight, and letter opener on his desk—that these routines showed up only occasionally and should have been used more often.

About the latter, I was wrong. For what this program was doing was finding a way to turn the conventional family situation formulas wrigglingly on their sides, and finding a way to let the comedy and the comedians operated on several levels. For Van Dyke, too great a reliance on "business" would have been a mistake, for he was finding ways to translate his abrupt, original, and understated sense of physical timing into verbal terms. And before this series was over, its cast and its writers had brought it to a kind of perfection, and Van Dyke himself had developed into a nearly perfect performer for TV series comedy.

Any family comedy series will eventually undertake a script featuring the embarrassed (and sometimes embarrassing) confrontation of father and son on the subject of the facts of life. But the Van Dyke show undertook it in a wild and mysterious episode which began with the premise that Rob Petrie, in good enlightened fashion, had already informed his son Richie on the matter. But a hurry-up and stilted meeting with the school "guidance counselor" reveals that Richie has been giving regular and quite outlandish sex lectures to his schoolmates. Perhaps Richie just got things a little mixed up the first time; there is nothing to do but stick to one's principles about sexual frankness for the modern child and tell him again. But Richie, told again, and told to keep these things to himself, nevertheless continues to beguile his classmates with bizarre accounts of where babies come from. Yanked away from his job at midday, father, with the school principal, the guidance man, and mother have a hasty, defensive, embarrassed meeting to get to the bottom of this unsettling matter. Richie soon resolves the problem of course. "Well they *like* to hear about that stuff, Dad. And you told me not to tell them what you told me, so I just made up something."

Even if a show does offer switches on the conventional, its performances, its presentation, its timing are still all-important. Briefly, the set-up has Rob Petrie writing for comedian Allan Brady, who is tyrannical, talented, egomaniacal, and when he wants to be, charming. The comedian's producer is browbeaten third banana Mel Cooley (Richard Deacon), actually his brother-in-law, who in turn is (or tries very hard to be) a tyrant to everyone else. Petrie's second bananas are his writing team. Morey Amsterdam (Buddy) can deliver a cuttingly insulting one-liner with such friendly innocence, disguised hostility, and quick timing as to make it very funny. The other member of the team, Rose Marie (Sally), is conceived as the stock middle-aged spinster who chases her men with just enough bright hostility-of-the-tongue to convince us that she doesn't really want one. Then there is Ann Morgan Guilbert as Millie, wife of the Petrie's next door neighbor, a kind of dumb brunette and foil to Laura Petrie—unself-conscious, brash, and unaware where Laura is conventional and reticent.

And so the show evolved a style with something for almost everyone, putting to good use the snappy, lowbrow one-liners of Amsterdam (a one-man joke file) at one end, and the relatively sophisticated exchanges of Van Dyke and Miss Moore at the other. Containing it all was that family format, which no doubt reassured the middle-aged but which, as I say, sometimes turned their values sideways without quite letting them know it. In the process, Van Dyke refined the kind of timing he had used in his older bits like the recovering drunk, and applied it to his speeches, underlining them with an agile and lanky body english, with his own "takes" and bits of business. In short, Van Dyke made him a TV comedian who registered primarily with his voice—we *hear* it first on TV and see it second; that (the exact opposite of the movies) is the nature of the medium. One tribute to his original craftsmanship, and that of Miss Moore, came about when the two of them were able to sustain a very funny half hour out of nothing more promising than the idea that Laura's big toe got stuck in the drain of an off-screen hotel bathtub.

The show's scripts, or the versions of them that reached the screen, were undoubtedly cooperative efforts among the writers, directors, and performers themselves. Morey Amsterdam surely provided his own gags (one can imagine his scripts consisting of blank spaces between cues) and probably some of Richard Deacon's and Rose Marie's too. Lines, business, all were tried out and tested during rehearsal, all were subject to revision and replacement.

In my own current memory, there is one show which especially stands out, but that is perhaps because I have recently seen it again. I am not sure of its title, but to me it is the "Pygmalion Dog" episode.

Allan Brady's TV guest was to be a canine film star, and Van Dyke, Amsterdam, and Rose Marie are stuck to know what kind of sketches to provide for a dog. Someone comes up with the idea of having Brady do the *Pygmalion–My Fair Lady* bit, taking a disheveled canine and turning him into Rin Tin Tin or Lassie or whoever it was. "Well, somebody go out and get a rotten-looking dog." And someone does, leading in the mangiest, foulest, dirtiest, ugliest, most doleful-looking hound that ever the ways of nature and the skills of the makeup man produced.

Various gags developed, including Van Dyke's slight but sneezing allergy to dogs and the fact that at quitting time he and the mutt find they are the last two left in the office. Therefore, the mutt has to be taken to the Petrie household. Appropriately, it is raining hard. Laura is of course inadequately prepared for the appearance and character of the beast, not to mention the strong smell of wet dog fur. And her efforts to carry off her dismay gracefully soon give way. While toweling off the animal, she encounters the last straw and looks up, repelled, "Rob, has he been eating onions?" (It develops he has; he had snatched Buddy's hamburger during lunch.) Even Richie, expected (his parents fear) to take the pathetic mutt to his heart in traditional boy fashion, declares him "funny looking." The hound ends up in a chichi poodle parlor where Billy de Wolfe, in a nice burlesque of a hairdresser but without too much obvious camp, agrees to take him over.

The Van Dyke show did a rare thing for a television series; it quit while it was ahead. It quit particularly because Van Dyke wants to make movies, and one wonders how he will fare. Clearly he is enough of an actor to be funny in a comic situation. But he has become enough of a television actor to sublimate all of his other skills—timing, movement, business—to his primary skill, that of reading comic lines with a secret combination of earnestness and lightness that makes them both dramatically convincing and comically effective. I doubt if he will do well in films unless he can modify his style. And I think of Jackie Gleason, a comedian who returned to TV after an unsuccessful attempt to work in films. I think also of Efrem Zimbalist, Jr., who became one of the best actors ever developed by TV and an ideal dramatic series lead. When he went into movies, he did not succeed, and he too headed back to television, his true medium.

(*Evergreen Review*, February 1967)

Clownish TV:
II—Red Skelton

Red Skelton is now in his—what is it? fourteenth?—year on TV. One could imagine that he may easily have twice that many more.

It would probably have occurred to no one to call Red Skelton one of our best comedians until Groucho Marx did so in a book of his a few years ago. It would not yet have occurred to anyone to call Skelton even a very good comedian if Skelton and television had not met. For television has revealed Skelton's resources, and he in turn has revealed its, and this mutual arrangement makes Skelton one of our most brilliant comic performers, as well as one of our most extremely uneven ones.

In this Red Skelton is like a number of comics—Milton Berle and Lucille Ball, for examples—who found their medium, and hence the real nature of their talents, on television. Take Miss Ball: she spent several years being a knowing, wisecracking cookie (a rival studio's answer to Joan Blondell and Glenda Farrell), and a few more years as the voluptuous lead in technicolor musicals (a rival studio's answer to anybody's glamour girl). Then she and TV discovered that she is a fine mugging, eyepopping female clown, another proposition altogether. So Red Skelton spent years trying to be a nervous, smart-aleck comic, always ready with a quick, smirking quip (a rival studio's answer to Bob Hope). But Red Skelton wasn't very good with the wise-guy retorts in the movies, and he was only a little better doing his own sketches with his own variety of characters on radio. For at his best Red Skelton is a sloppy, rubber-faced, heavy-limbed, low-comedy buffoon.

The stand-up monologue with which Skelton usually opens his TV show still isn't much, and the comedian himself often seems to harbor a bit of suppressed embarrassment at standing there "in one" with the camera glued to his face as he delivers some fairly characterless jokes. But his embarrassment quickly leaves him for a performer's zeal if he finds his way into a really absurd story, or if he can break off the whole thing for a bit of pantomime.

The abiding quality of a Skelton show is its casualness. And this casualness both discovers Red Skelton as a gawky, spontaneous clown and gives one insight into the wide casualness which television can assume—a casualness which would quickly become unbearable on stage and would be unthinkable in a theatrical film.

Watching a Skelton show is often like watching a rehearsal for a show that may turn out to be good or may turn out to be terrible—and there is sometimes just no telling which. It has been said that one reason for Skelton's durability on TV is that he is likely to play a different character on each of six shows, and thus he doesn't wear out his welcome by repetition. One week he is Dead Eye, the inept Western badman; the next, Klem Kadiddlehopper, the apparently mindless rube; the next, George Appleby, the henpecked husband; next, San Fernando Red, the Southern con man; next, Freddie Freeloader the self-satisfied circus clown-bum; and from time to time, he even undertakes to put on short velvet pants and be "the mean widdle kid." The truth is that there's not all that much difference among his characters—or rather that after a scene or so (and no matter what kind of plot his writers have provided him with) Skelton becomes Skelton, apparently clumsy but strangely adept, enjoying himself enormously, ad-libbing outrageously, trying good naturedly to break up the rest of the cast, and, when the spirit moves him, stopping everything in order to rattle off some favorite jokes on the subject at hand. He may also give casual instructions to the actors, the cameramen, and the fellow holding the cue cards. It usually works out very well—or at least it works better than sticking strictly to the script might have worked. And when the moon is right, it works brilliantly. But in any case the point is that it *can* work and that it can reveal Skelton.

True, the Skelton break-up is not always of his own making. Having seen how effective the device can be in Skelton's own ad-libbing, one of his writers may simply put an aside or a crack about cue cards into the script if he thinks of a good one. The practice probably started with Milton Berle, to be sure, and it has reached some kind of crisis on the Dean Martin variety show, where apparent ad-libs, asides, fumbles, misreadings, and break-ups come thick, fast, and effective, and, we are told, are not always so spontaneous as they may seem. (Even in the relaxed atmosphere of the late hour, ad-lib TV talk and variety shows like Johnny Carson's or Merv Griffin's, there is more planning than may appear to the audience at home.) But the point, after all, is how agreeable and even entertaining such extreme informality can be on television in contrast to any other medium, whether that informality is always spur-of-the-moment or not.

Not that Skelton's writers let him down, just that they know their place and keep it. And nothing seems to delight him more than when they provide him with a fast-moving, conventionally-based plot line that in its details is put together as wildly illogically as possible. Of course he likes it when they give him a good eye-rolling comeback like:

WOMAN: Do you want me to wind up an old maid?
RED: I hope not because when they unwind they're murder!

Or the writers may even give Skelton a straight line so long as the gag which follows it pleases him. Red observes an extreme January-June marriage:

RED: My goodness! She's so young I don't see how you got a license.
TOTTERING OLD MAN: We couldn't, young fella. So far we've been getting by on a learner's permit.

The more his material hints at raunchiness, the better Skelton likes it (and one of his ways of breaking up the cast is to allude to the blue material he notoriously—and innocently—ad-libs during rehearsals):

CUSTOMS INSPECTOR: So you aren't trying to smuggle anything in, eh? What have you got under your arm?
RED: Hair.

So far, Skelton is defying no law of TV aesthetics, for his costumes; his shuffling, knock-kneed gait; his plastic face; his stringy, unkempt hair; his heavy, gangling arms; and all the dismantling props he picks up are merely complements to a monologue of Skelton's own, which acknowledges the presence of Richard Skelton himself, the clown Red he plays onstage, the audience he likes to please (but very much on his own terms), and some other likable actors who must, for the sake of a plot line, interrupt him from time to time with speeches of their own. But there is one aspect, perhaps the greatest aspect, of Red Skelton which would seem to break the laws of TV, and that is Skelton the pantomimist.

Almost anyone who has ever seen Marcel Marceau at work on a stage, critics included, will readily declare that here is a great pantomimist. But until 1965, few people would have called Skelton a great one. Then Marceau himself did, and at Marceau's instigation, he and Skelton did two all-pantomime hours together on TV. And irony of ironies, on the first (and much the best) of them, Skelton outshone Marceau.

Marceau moves us strongly and often deeply, but he is, of course, a studied perfectionist, and each of his routines is planned and timed probably down to the last quarter-inch movement of a finger. He plays his Bip-Everyman to the world at large, or, in more practical terms, to the theatrical house at large. His art involves the basic paradox that, although the world of Bip is small, the experiences of Bip are universal, and the gestures of Marceau, for all their delicacy, are broad and even sweeping in their effect. Skelton in his way is no less skillful than Marceau, but his art is far more intimate. And in such an intimacy there exists a possibility for improvisation, for departure from a routine under the inspiration of the moment. Skelton works by mutual agreement between Red and you. And between two people there can exist a casualness that would be unwieldy for Marceau, who works between an impersonal Everyman and the large gathering which fills the seats from orchestra to gallery. And this means also that Skelton projected to the camera and through it to the living room screen. Marceau, alas, did not.

Of course Skelton's pantomime is not the refined, high comedy of which Marceau is the greatest living practitioner. Skelton, once more, is the big-footed, flabby-mouthed buffoon, and he may seem willing to pull any kind of low gag for a quick laugh when one first encounters his work. But one is soon aware that Skelton the pantomimist lives in a world where every absent-mindedness, every physical awkwardness and ineptness, and every mechanical breakdown that *can* happen eventually *will* happen. A ballerina (Skelton) will always find that her partner (also and simultaneously Skelton) just won't let go of her after a lift. The middle-aged do-it-yourselfer is foreordained to bang his thumb, saw off a couple of his fingers, and have the head of his hammer fly loose and smash through a nearby window. And always such conventional mishaps are greeted with such complete surprise and total innocence on Skelton's part that he can make each of them freshly hilarious.

One of his funniest routines has to do with a doctor performing extended surgery. But as Skelton sees it, such a doctor (a) is amiably drunk, (b) sneezes all over his hands after a lengthy scrubbing-up (the antiseptic powder gets up his nose, of course), (c) is surprised and alarmed by what he sees after he cuts a wide swath across the patient's abdominal cavity and peers inside, (d) picks up a pulsating heart and places it on a shelf behind him for sake-keeping while he yanks out, looks suspiciously at, and scornfully tosses aside one vital organ after another, and (e) sews the patient up, getting quite carried away with his fancy stitchwork, but of course forgetting to replace the now faintly pulsating heart.

But Skelton's abilities as a mime are not confined to such comic ineptitudes and low buffooneries, for his famous "Old Man Watching a Parade" is as good an example of brave and unsentimental pathos as contemporary comedy can offer.

I don't think that in such effective work in pantomime Skelton refutes the principle of TV as a visual medium only secondarily. The effective speeches of, say, Paul Muni as Zola, or Laurence Olivier as Henry V (succeeding where other actors—George Arless particularly—had failed in making us truly *listen* to a film) do not refute the principle that the movies are primarily visual. Such achievements demonstrate that a medium can be more complex than anyone had previously realized. (And it is not without meaning that Skelton keeps an off-screen sound-effects man busy during his routines.) At the same time, Skelton's way with mime is particularly adapted to TV as a medium that can be both highly personal and highly informal.*

In all this, Skelton works not so much out of his experience in movies, as out of his experience in radio and uses the pacing he had learned to use on his radio shows.

Skelton's raucous pantomimes, as apparently casual, offhanded, and semi-improvisational as his sketches, are also as intimate. Skelton learned that, much as acting on film has to be taken down several emotional notches from acting on stage, TV performing has to come down several notches still. It has to be much more direct and intimate. In movies, the camera crowds in on the actor, but the audience may still be spread out over a city block. On TV, not only is the

---

* Admittedly, as TV screens have become larger, the visual effectiveness of TV has become greater. And as they become larger still, the balance of aural-to-visual will undoubtedly change.

camera four feet off, the audience virtually sits in your lap. And on radio, the mike had been in your face and the listener a few feet away.

True, one needs to watch a great deal of Skelton to get the best of Skelton. But the best, I think, is worth the effort. And I am grateful that middle-grade Skelton is good and that, as he offers it, his work is so casually a matter between him and me that criticism becomes outright unkindness.

(*Evergreen Review,* April 1967)

Cops and Robbers

If the United States could be said to have a national literature, it is crime melodrama. By that I do not mean that it was an American, Edgar Allan Poe, who established the *genre* of the detective "puzzle" story. I mean that the bulk of our fiction is cops and robbers stories. I mean that an average American novel, film, play, or TV show that offers crime melodrama is more than likely to be competently done. I mean that even a poor one can be interesting and entertaining in one way or another. I mean also that the Western badman and the big city gangster are virtually interchangeable types. I mean, further, that an outstanding directoral reputation in films can be built on a well-done crime movie—no one was taken aback when John Huston, having done a good job with *The Maltese Falcon,* moved on to the likes of *Moby-Dick* and the Holy Writ.

Our more moralistic observers have told us that our crime fiction glorifies crime and glorifies the criminal. The charge is often glibly delivered, but it cannot be glibly answered. But by way of an answer, one could reasonably postulate that the criminal is often a potentially complex being and hence he is a dramatically interesting being.

Without attempting to exhaust the subject, let us try a somewhat different tack. Symbolically, the criminal of fiction has made the first decision of masculinity—that is, he knows what he wants. It may be that he has made this decision wrongly, but that is not the point. He has not, however, made the first decision of maturity, which is to acknowledge and accept the responsibilities and the consequences that go with knowing what he wants. Thus it is possible for the average gangster film to lull its audience into a kind of wish-fulfillment. Get them to identify with a protagonist who takes what he wants, guiltlessly, ruthlessly, and regardless—that is, regardless of everything except the required pseudo-moral ending in which that protagonist is rather arbitrarily arrested or shot down or whatever.

However, the gangster protagonist is potentially a tragic being. If he discovers, usually by falling in love, that he does have within him some regard for the desires and needs of others, it usually comes too late for him, and that essential discovery dooms him. Thus James Cagney's Chicago hoodlum in *Public Enemy*. Or Humphrey Bogart's prison escapee in *High Sierra*. Characters like these in our crime dramas and in some of our Westerns probably bring us as close to true tragedy as American dramatic fiction has ever moved.

As crime drama is a staple in our films and our literature, so it is on television. In the late forties and early fifties, at a time when TV drama was still dominated by New York-based "live" broadcasts, along came a Hollywood-filmed series called "Dragnet" which proved an instant success and had a several years' run before that success waned. And, wonder of wonders, "Dragnet" tottered back during the 1966-67 season like an aging grandparent, a quick replacement for a mid-season flop. And "Dragnet" gathered the audience ratings once again. But "Dragnet" was a prolific grandparent to sometimes brighter children. "Gunsmoke" began, first on radio, as a kind of Dragnet-on-the-range. And "Medic" began as a Dragnet-in-a-hospital, hence "Have Gun" is a step-child. So is the current "Jack Webb's True." I don't know that it has occurred to anyone to say so, but the "procedural" school of whodunits (Ed McBain, et al.) owes its impetus to "Dragnet."

There were plenty of Hollywood-produced half-hour crime shows on TV when "Dragnet" originally appeared, of course. Most of them were produced roughly along the lines of "quickie" grade B and C crime films that our film studios had been turning out for over thirty-five years. But "Dragnet" was different and, for all its shortcomings—apparent and appalling though they be—I think that its success was deserved and that its arrival was a pivotal event in TV history.

It is hard for us to realize it, even when we know the facts, but "Dragnet," this most stylized and artificial of all TV series, began as an effort at realism. Some official or officials of the Los Angeles Police Department chided the procedures of radio crime and private-eye shows at the hokum and unreality of criminal actions and the police work involved in their scripts, and, with a generous gesture toward the Department's files, declared that the facts of police procedures contained the stuff of better drama.

Thus, "Dragnet" with Jack Webb, beginning as a radio show. Soon its success on radio meant its transference to filmed TV. And there the stylization that had developed on the show became more apparent. Actors involved in a well-written vignette revealing the vagaries of human character seldom raised their voices or their tempers above the impatient monotone of Sgt. Joe Friday. Nearly everyone talked like everyone else, which is to say that nearly everyone took on the anti-style of Jack Webb. Preachiness intruded. And dramaturgy sometimes flew out the window.

There was an early show—there was probably more than one—in which an important piece of information was withheld from the audience until the climax. For suspense, you may say. Apparently not, because the withholding contributed no suspense. It was as though the writer simply forgot to mention that an important clue had been found at the scene of a crime, or the director had forgotten to shoot the lines that covered the event, or the film cutter had somehow removed them. And nobody was bothered by the result.

Sometimes, in the attempt to find an appropriate TV style, there was a pompously irrelevant sort of "visual" artiness: when a suspect finally confesses at the end of a tense interrogation, the camera rapidly closes in on a half-eaten apple left over from Friday's lunch. In a sense, Stan Freeberg's parodies of "Dragnet" weren't needed— the series had provided its own.

Well, then, how could such a show be bearable, and if bearable, how could it be important? "Dragnet" was bearable, first of all, in the way that most cops and robbers shows are bearable: we keep watching to find out how it is going to turn out. It was bearable also because it frequently did give us, in the writing if not always in the performing, brief flashes of interestingly observed characters—eccentric, if you will, but never deliberately bizarre—characters who are, we sense, the almost unique result of twentieth-century urban American life. And the scripts abounded in them.

And "Dragnet" was important because it was one of the first television shows to attempt a true television style. But, for the most part, a "Dragnet" episode is a series of pictures of people talking to each other. It was a photographed radio script, then. And thereby "Dragnet" demonstrates the crucial aesthetic fact of TV.

And the stylized monotone acting? Well, I think that there Jack Webb, tentatively, even awkwardly, was also getting at a further essential of TV technique. Just as film acting has to be less broad, more understated than stage acting, so acting on TV must be more underplayed. To put it in obvious terms, a stage actor is projecting at least to the back of the orchestra, if not the center of the top gallery. On the movie screen, the camera is perhaps only a few feet away, but the audience, on the other hand, might be a quarter of a mile away. On TV, the camera is in tight, and the audience is six feet away. When a performer is virtually sitting in his viewer's lap, he must bring his performance down even more than for cinema. Webb and his fellows sensed all this.*

---

* I might remark here on the extent to which series characters gradually take on the characteristics of the actors who portray them. When Harry Morgan first joined M*A*S*H, his own interest in horses was deliberately made one of Colonel Potter's interests. But earlier Corporal Klinger, obviously first conceived as Jewish-American,

Perhaps it is this necessity for an intimate acting style—a necessity which Webb and "Dragnet" never learned to handle perfectly—that has led one commentator to call TV a "cool" medium. But it seems to me that anyone who thinks that so eminently successful a TV performer as Lucille Ball is "cool" is sadly deluded. Nor can I concede that the medium is the message. The "message" in drama is constant, and by and large the medium doesn't determine that message. Each medium—stage, film, TV—finds its own way of presenting it, of making it effective. And television's way is primarily auditory, secondarily visual, and histrionically underplayed.

To go back to the subject of cops and robbers, one can find as many good examples of it as he needs in the TV fare of any given season. When "Naked City" was on, the task was perhaps easier and more rewarding. But to draw my examples from the 1966-67 season when "Dragnet" was reintroduced, I would cite "The F.B.I." and "Felony Squad."

It is a bit difficult to discuss "The F.B.I." series without acknowledging that any of its scripts that dealt with spying, espionage, or anti-Communism was apt to present our national propagandistic self-righteousness at its worst. But when the series dealt with domestic crimes and domestic criminals, its approach was often a good deal more complex, and it does not really matter if the points of departure for such complexity were sometimes hackneyed. The playlets usually left the agents and policemen alone; at the end they were what they were in the beginning, what they were last week, and what they will be next week. But not the criminals and their victims.

The father of a kidnap victim discovers by nearly losing his son not that he loved him but that he had always treated the young man with a patronizing lack of confidence. The son makes a similar and

---

was made Lebanese because Jamie Farr, who plays him, is Lebanese. And Mannix, obviously first conceived as an Irish-American private detective, gradually became Armenian because Mike Connors is of Armenian extraction. Mannix was even given a family and some adventures in the California Armenian community. And on "Alice," the heroine's supposed Italian background is often confounded by her knowing references to such things as bar mitzvahs, which surely come from actress Linda Lavin's own heritage. Et cetera.

opposite self-discovery. The morally unaware girlfriend of a young hoodlum realizes that she can want him arrested and stopped, and yet still love him. A pair of lethargic and cynical young people are stimulated into eager activity, but into ultimate self-destruction, when the temptation to undertake a perfect robbery crosses their path.

The final result is unsatisfactory of course. We are left with the impression that one lives his life either in the rather complacent moral conformity represented by the F.B.I. men, or that one is awakened to a more complex and interesting existence only through criminal activity and potential self-destruction.

I am aware that it is possible to dismiss this show's efforts at characterization, and at dramatic action, as merely the currently fashionable paraphernalia of crime fiction. But I think that such efforts are not "merely" anything, least of all fashionable. Admittedly, "Naked City" handled the implications of character and morality more satisfactorily on the whole than does "The F.B.I.," and anything another series conceived along similar lines will probably be derivative of "Naked City's" achievements, at least for the next few years.

"Felony Squad" succeeds almost as "Dragnet" fails; it succeeds in seeming real, where the older series has come to seem artificial. It presents us with fewer quirks of behavior on the part of the citizenry, and sometimes tries to look deeper into the character of the cops.

I will cite one "Felony Squad" episode called "Break Out," written by Barry Oringer and directed by Larry Peerce. On the face of it, the script offered us nothing more than the unexpected capture of a boss hoodlum, and a tricky and complex intrigue with which his henchmen attempted to free him from an overnight lock-up at the station house. At that level there was plenty of suspense and enough action.

But in its execution, we got a great deal more. When the hoodlum was first brought in, for example, the press leaped upon him and upon his captors outside the police station with an eagerness that quickly became arrogance. And when the reporters were given little information by the arresting officers (who did not want to *invite* a break-out), they behaved with an arrogance that approached viciousness. A viewer came away feeling he knew something important about what it is like to be a cop.

The show's final scene was staged in the police parking garage, with the criminals trying to escape in their usual limousine, and the cops, in a kind of frantic automotive chess game, trying to block their way by moving police cars in their path without getting shot down or run over in the process. It was an exciting sequence, not only because the material was potentially exciting, but also because the staging handled it well and because whoever edited it was able, partly by crosscutting just enough dialogue, to let the cars symbolize the mortal determination of both the cops and the robbers.

One major asset of "Felony Squad" is Howard Duff as Sgt. Sam Stone. He is a better actor at middle age than he was as a juvenile—or perhaps I mean that he is a better television actor than film actor, and surely his background as a radio actor serves him well on TV. He helps the scripts and the direction tell us what it must feel like to be a policeman, and a morally aware policeman at that. He also makes us wonder why anyone would undertake to be a cop, and yet know why Sam Stone probably could not undertake to be anything else.

How can a series do this well if it is turned out week after week, thirty times a year? It stands to reason, *some* sort of reason, that nothing turned out this fast and this often could be very good. But whatever reason may have to say about it, experience shows us that about a third of these cops and robbers shows *are* good. Television is a medium for those who can work well fast. Anyone who works well only when he works slowly had better stay out of it. And any man—actor, writer, director, producer, technician—who does work well when he works fast may find TV the only challenge.

(Unpublished, 1968)

Take What You Can Get

Last month, when Channel 13 and PBS presented a prestigious anthology of dramatic sketches by William Saroyan, most of our reviewers gave the show their obedient, even reverent, attention.

None of our reviewers, so far as I know, paid any attention to an episode of Raymond Burr's "Ironside" series that ran the same month, called "Too Many Victims." But they might have, for Burr did a truly remarkable acting job on that show.

The situation was unremarkable, to be sure. Burr, as Chief Robert Ironside, had to turn in another policeman, a lifelong and otherwise respected friend who had deliberately framed a dope pusher they both knew to be guilty. Using nothing but his face, arms and voice (Ironside is a cop crippled in the line of duty, and Burr plays him in a wheelchair), Burr offered in several scenes a rare insight into the nature of human feeling and human responsiveness when they are at war with human reason.

Burr is often that good on his series; to those of us who watched him stalk his way through all those "Perry Mason" episodes, it is sometimes hard to believe how good he is. Once, handed the probably inevitable script in which a famous surgeon raised the possibility that the use of Ironside's legs might be restored by an operation,

Burr had a superb actor's moment. It is possible to describe it only by saying that he actually was a once-active man who had lost the use of his legs years before, and was given a tortured moment of hope as he felt a faint sensation in one of them.

If anyone wishes to describe what Burr shows us so often on this program—the fundamental conflict between emotion and reason in a man of sensitivity and intelligence—if anyone wants to describe this as merely "escapist entertainment," he is welcome to. But I think he will be very wrong to do so.

The star of the Saroyan playlets on Channel 13, to get back to them for a moment, was Pat Hingle, and he came in for his share of praise in the notices. Last season, however, on a "Medical Center" episode, Hingle brought off a well-conceived and beautifully acted scene in which a basically inarticulate, successful businessman tried to explain himself, across the generation gap, to an alienated daughter. I did not read much about that accomplishment in the TV columns.

I do not read very much, either, about Carol Burnett's sketches in television reviews. Any week she is likely to do a spoof of 1930s Hollywood musicals that will make the successful Off-Broadway *Dames at Sea* look rather pale and self-indulgent by comparison. A season or so back, there was even a masterful, hilarious burlesque in which Miss Burnett, assisted by Mel Tormé, managed a double comic perspective by giving us a 1930s Hollywood version of turn-of-the-century backstage life.

A successful TV producer candidly confessed that he feels he has done well if a third of the shows he offers in a given season turn out to be good ones. But that seems to me an absurdly high number. Suppose a drama critic declared that a third of the shows that open on or off Broadway were really worth seeing. Or suppose some book reviewer announced that a third of all the novels written in a given year were eminently worth reading. Either one might lose his job for simple-minded incompetence, or at least set his readers' teeth on edge for anything he might write in the future.

Our play and book reviewers, our music and ballet critics, do not remind us that most of what they encounter in a year is not very good, because they assume we know that goes without saying, yet somehow, with television, all that does not go without saying. And it seems to me also that some good things on television pass by with little critical attention.

Television is popular art. Any writer who does not have a feel for popular art and its wellsprings had probably better leave it alone. As the late Erwin Panofsky, a great art historian and a very wise man, once wrote, popular art has at its base a frequently crude sense of justice, a sentimentality, an almost primordial instinct for violence, mild pornography, and a crude sense of humor.*

From this base, it is only a question of what the individual artist can do with the idiom in which he works. And I venture to say that the man who does not understand that will no more truly understand Shakespeare than he will understand Louis Armstrong, no more truly appreciate Rabelais than he will Duke Ellington, Dickens than he will Fred Astaire.

If a man has no feel for popular art, he will not know that mystery writer John D. MacDonald is one of our very best writers, and that the shortcomings of his remarkable novels *The End of the Night* and *A Key to the Suite* are probably neither more nor less than those of John Updike's best work, only perhaps different in kind.

* To clarify a bit, I believe that the American arts and the American genres of art (jazz, the movies, comic strips, detective-suspense tales, *et al.*) can best be understood within the traditions of "popular art," or at least European "popular art" (Giotto, Dürer, Shakespeare, Dickens, Bizet). However, in this country, and in this century, a distinction between "popular" art and "fine" or "aristocratic" art grows more and more useless, and the increasing use of terms like "popular culture" has become a way of patronizing some of our most unpretentious but creative artists (D. W. Griffith, Buster Keaton, Fred Astaire, Duke Ellington—the list is a distinguished one). I am also convinced that the "popular culture" movement among our academics has an implied condescension. In any case, the study of "popular culture" seems to find equal interest in Mickey Spillaine and Dashiell Hammett; Roger Corman and John Ford; Cher Bono and Sarah Vaughan; Norman Rockwell and E. C. Segar; "Bonanza" and "Gunsmoke." So perhaps it is yet another way that our intellectuals have found to avoid the task of sifting and evaluating the work of our most original artists and our most characteristic contemporary art.

So one takes television for what he can find and, if possible, offers it his patient love. He enjoys the inspired whimsical zaniness (a tough combination, that) of "Green Acres" on a good night. He loves the incredible gradations of feeling that flow through Dean Martin's very presence on the small screen; it is as if Martin were saying, "I don't care much about this script, but I think we can have a heck of a good time together, so let's do it!" (Dean Martin is quite possibly the greatest straight-man in human history—anyone who can play a scene with Bob Newhart, who is essentially a monologist, and make him even funnier. . . .)

Those of us who love television know that the starkly tragic, pastorally innocent, righteously melodramatic, bickeringly comic world of "Gunsmoke" has offered some of our most deservedly popular recent drama. We know that Lucy is (or can be) a great clown. And we know that Flip Wilson is fast becoming one; we do not need future critics or historians to tell us that. And we know about the exceptional and remarkable versatile character actors who work on television, men like Charles Aidman, Steve Ihnart, Edward Asner—but to mention some of them is to neglect others whom one should not neglect.

Those of us who watched Richard Chamberlain evolve from a somewhat insecure, aw-shucks, stone-kicker to a confident young actor on "Dr. Kildare" are not at all surprised that he has done an impressive *Hamlet*.* In so doing, it is surely worth pointing out, this "television actor" has undertaken what our most praised and talented stage and screen actors of this century, Burgess Meredith and Franchot Tone through Marlon Brando and, thus far Stacy Keach, have not undertaken.

I hope one of our museums has a print of Chamberlain's *Hamlet*, with all its violence and sex. I also hope it has (at random) the best of Phil Silvers's "Sergeant Bilko"; Ernie Kovacs's gorilla-suited corps de ballet doing "Swan Lake"; the early "Defenders" show

* A *Hamlet* of sustained petulance and reluctance is a valid *Hamlet*, it seems to me.

which raised, and *faced,* the moral consequences of a violent act in a manner that I think Ibsen would have understood but Arthur Miller would not. I hope the museum has the best dances Ernest Flatt staged for the old "Garry Moore Show"; Fritz Weaver's fine, tragic performance in Larry Cohen's well-conceived script, "Medal for a Turned Coat," on the "Espionage" series; the best of Red Skelton, particularly his first all-pantomime show with Marcel Marceau; the best of "The Honeymooners"—and, yes, the best of "Rocky, the Flying Squirrel." Whatever museum has all that will have a record of some of the best theater our time and our country have produced.

(*New York Times,* December 20, 1970)

IV

Sitcoms and Cities in Trouble

Comment on TV in the 1970s

The Progeny of Archie Bunker

There are those who still object to "All in the Family" and its successor, "Archie Bunker's Place," as fostering prejudice and bigotry. Probably they would prefer their bigots to be readily identifiable villains in traditional melodramas rather than occasional buffoons in TV comedies.

The trouble with putting social messages in melodramas is that melodrama is patently prejudiced by its nature, and the audiences can't possibly identify with its villain-bigots. *Dark at the Top of the Stairs* or *Separate Tables* would provide two relatively recent examples wherein audiences could come away feeling they have had an uplifting human experience, gained great understanding, simply by despising an obvious badguy or two. Never, of course, having been invited to examine their own biases or their own potential for bigotry.

"All in the Family" invites us to laugh at and ridicule Archie's bigotry, while inviting us also to examine his blundering, foolish, and pompous humanity as perhaps a bit like our own, and (we hope) as redeemable as our own. And Archie Bunker as developed by the show's writers and by Carroll O'Connor is, for all his surface simplicity, a complex comic creation.

I have earlier expressed apprehension lest Archie be made too lovable. On the new version of the show he has indeed become more tolerant, or at least a bit more aware of his rampant intolerance. He is even raising his adopted daughter as a Jew. But as he becomes more likable and a bit more vulnerable, so do his ignorance and prejudice seem more willful and more absurd—and such is the nature of comedy.

But if Archie Bunker invites us to test ourselves, so also did his son-in-law Mike, for Mike's tolerance was often intolerant, shallow, and belligerent. It tested the writers too, I expect, for their sympathies were surely with Mike's ideas if not always with his eager rush to the side of righteousness. In any case, we all understood that the truly tolerant member of the family was Edith, a being of infinite good will even when the innocence that so often protected her was shattered.

And "All in the Family" gave us the episode of the death of the old man, and "Archie Bunker's Place" gave us the episode of Edith's death, and any series which can deal with death with reverence and stinging pathos, and still make us laugh with and at its characters, has given us excellence.

The effect of Archie's presence on TV is not so much in the glib, snappy-jokes-and-social-problems formulations of "One Day at a Time" or "Good Times." It is that the blue-collar Bunkers have brought back the possibility of true low comedy to the tube. And "Laverne and Shirley," that most maligned of all recent shows, has brought us low comedy of a most unusual kind.

First of all, it is surely worth remarking that Laverne and Shirley, in the persons of Penny Marshall and Cindy Williams, have given us our first successful female comedy team, and there have been earlier efforts, particularly in the early days of sound film. If Laverne and Shirley do not quite reach the level of Laurel and Hardy, well, neither do they offer the drone competence of an Abbott and Costello. It is also worth remarking that in Lennie and Squiggy (David Lander and Michael McKean) we have the only team of *zani* we have ever had—no straightmen, just two appealing crazies.

There was an episode which had Laverne convinced she was pregnant after she passed out (from one beer!) during a ride through the Tunnel of Love. It included a scene of fine outrage, anger, wrath, sympathy, love, and support—all at once and superbly executed by Phil Foster as her father. And it had Lennie's cautious but genuine offer of marriage:

LENNIE: You see, Squig and me, we drew straws . . .
LAVERNE: And you lost.
LENNIE *(gently)*: Oh, no! I won.

Sure, raucous low comedy is often a matter of taste, and it always involves the risks of taste. There was an episode in which the two protagonists helped stage a wedding in a local black Baptist church (it was somehow the only place available) and ended up joining the choir, and that segment seemed to me to cross the line where a broad burlesque became (in this case) racial insult.

In any case, to enjoy such low comedy at its best, one simply has to have a taste for farce and for clowning . . . and, well . . . alas for those poor unfortunates who do not have such a taste when they are faced with Laverne and Shirley.

As I write, the most recently arrived blue-collar comedy on TV is "Flo," and "Flo" (played by Polly Holiday) seems promising— and that strikes me as most unexpected.

"Flo" is nominally a spin-off of "Alice," a series which seems worth watching only to see what competent players can do with scripts that might have been put together by a committee of showbiz wiseguys.

"Flo" has its good-old-gal heroine as proprietress of a small Texas crossroads bar, with a reluctant male staff, a richly addle-headed mother, a determinedly prim sister, and a potential for some townspeople *versus* truckers set-tos. All of which would seem to open the door for some traditional New York-Hollywood condescension toward all those redneck yokels, colorful though they may be. No, that door has so far stayed closed, and the series has treated its characters with a genuine delight in their ways and occasionally an agreeable pathos. The episode which opposed the realities of a horrendous Thanksgiving family reunion to the coziness of a Norman Rockwell painting prominently hanging on the dining room wall had a lot going for it.

Non-Violent Saturday

"Too much violence is not good for children," an eminent American pediatrician said about TV programming on a talk show recently. "We know that for a scientific fact."

We do not know that for a scientific *statement* surely, anything but. But it contains an interesting slip, for if too much violence is not good, so necessarily is too little.

I introduce the quotation and raise the subject in order to deplore both the blandness and the moral hypocrisy that well-intentioned, do-good meddling and pressure have wrought in Saturday morning children's TV programming during the last decade or so.

The blandness and dullness and pointlessness are patent and probably not worth discussing—except perhaps to point out that the writers now have few plot devices to fall back on, except the one that involves someone chasing someone, in order to . . . well, to scare him or something, it seems, for some vague but threatening reason.

The hypocrisy is perhaps worth discussion because time and again a powerful hero will pick up two antagonists by their collars or their necks and tell them in a determined voice that they should be reasonable, "talk out" and settle their differences without coming to blows. And the always implied but unspoken next sentence is, because if you don't, I'm gonna beat the hell out of both of you. And indeed, what else?

So for the child viewer, the presence of violence, and his own potential for it, instead of being dealt with overtly and resolved on its own terms, is repressed, off-camera—a vague, unspoken, amorphous threat, tied to a big-brother control.

To be sure, many idealistic Americans have not dealt with the presence of a necessary threat of violence in maintaining the public order, nor indeed with the darker side of everyman. And this is perhaps not the place for me to confront such issues, except to point out that they are being raised, and being piously and hypocritically repressed, on children's TV every Saturday.

One should be very careful in making claims for an absolute aesthetic freedom as the necessary progenitor of "better" art. The great Greek dramas were produced under a church and state control, and that control actually stimulated the dramatists—Euripides, overtly so. The Elizabethan dramatists worked in an atmosphere of state censorship from church and crown, Puritan disapproval of the frivolity of such entertainments, and the highbrow scorn of the educated followers of Sir Philip Sidney. And whereas one can say that our films were *different* once the "Hays Office" began to enforce its Code in the mid-1930s, can we really say that they became inferior?

So I cannot argue that, if it were not for the meddling of educators, parents' groups, and others, children's TV would be better. But I do think that it would be less forced into moral hypocrisy, and that the profoundly positive effect of an overt, symbolic, and purgative violence in children's entertainment (or anybody's entertainment) might find its way back into our lives.

Street Bizarre

The highest compliment one could pay "Kojak" and "Baretta" would be to compare them with "Naked City," and the comparison would be generally favorable.

Each of these shows gave us the image of a crumbling city, of capitulating members of its population, and of a small group of people assigned to bringing some order to its despair. "Kojak" and "Baretta" put strong personalities, bizarre, larger-than-life protagonists who care, and who, if they can guide no wars, can win some battles and win them with compassion. "Starsky & Hutch" began with roughly the same city-image but quickly and incongruously encouraged its protagonists to become smug, shallow, boys-will-be-boys clowns, and turned its city into a playground, where a self-righteous violence (borrowed, no doubt, from recent movies of Clint Eastwood and Charles Bronson) could flourish.

Kojak and Baretta; a clever Dapper Dan and a ragtag Trickster, carrying the load of everybody's failures, everybody's helplessness, everybody's neglect, everybody's wickedness.

Robert Blake's Baretta particularly managed a compelling combination of comic eccentricity and compassionate melodrama, and Blake could move from an interrupted seduction, to a distressful scene of an accidental shooting or a furious encounter with a killer, through a farcical encounter with Billy, his befuddled alcoholic landlord, or Rooster, the flamboyant pimp, with a kind of lowbrow grace that is surely unique in American drama. The episode involving Baretta's tribulations at the paws of a marijuana-sniffing dog, and unsympathetic bosses, sticks in the memory as a small black-comic masterpiece, and Blake's Baretta as a kind of flamboyant Huck Finn, functioning as a committed, big city, streetslum cop. Could any other medium but television have offered us his like?

"M*A*S*H" and the Balance

I find "M.A.S.H." one of the best shows on TV, and that surprises me. I found the movie version rather repellent and the critical reception that movie received puzzling. The film featured two callous, supercilious heroes who began their action by stealing a jeep, offering only condescension to the fearful corporal who was obviously going to get into trouble as a result; they lorded their superior intelligence and abilities over everyone; and they unashamedly used their surgical talents to gain favored treatment. The film set up Frank Burns as a religious fanatic and sneered at him. Henry Blake was an incompetent unit commander, and the film ridiculed him smugly. And its editors cut to shots of blood and gore whenever its script or continuity got into trouble. All in the name of making a not-at-all-disguised Vietnam protest.

Ah, but the TV series. With the temptations of the film lurking at hand, and the temptations of several horrendous fun-in-wartime sitcoms ("McHale's Navy," "Hogan's Heroes") just around the corner, "M.A.S.H." comes off as a strong comedy series with convincingly intelligent banter and insightful but restrained pathos.

True, in the very beginning Hawkeye (Alan Alda) was a bit smug and judgmental, and the quality lingers; Frank Burns (Larry Linville) kept a Bible by his bedside (but he didn't thump it); Radar (Gary Burghoff) was more overbearing than callow; and Trapper John (Wayne Rogers) was little more than Hawkeye's tent mate and drinking buddy. But most of that changed, and "M.A.S.H." discovered its humanity the way most of us do, by facing up to its shortcomings. A key event was the episode that had Alan Alda's father Robert playing a visiting surgeon who pointed out that Hawkeye was very hard on people and was capable of seizing upon their weaknesses.

One problem with discussing a series of such generally sustained quality is that merely pointing out outstanding shows is a difficult task. I will cite only the one that centered on a poker game and featured: a hip-talking Oriental; the area's dour psychiatrist; an accident to Radar with Colonel Blake as the unwilling father-surrogate pulled out of the game; the notorious "Whiplash Wang" faking another accident for retribution payments; Klinger's usual pushing for a discharge by sporting the latest Butterick fashions; and Burns's pompous statement that as a loyal American he couldn't operate on a

C.I.D. (read C.I.A.) man without another C.I.D. man on hand in case state secrets were mumbled under anesthesia. All this in one fugue-like script that was complex but not cluttered. As I remember, it was also in this episode that a berserk patient tried to shoot up the unit because he was "not going back."

Perhaps the achievement of the series that is most indicative of its quality is the maturity with which it has handled its cast changes and replacements: it has not merely sustained itself thereby, it has improved its quality. The careful original balance of characters and types has been repeatedly challenged but changes have only enhanced it. Henry Blake, the indecisive, bumbling unit commander, has been replaced by Sherman Potter, who is experienced, mature, insightful, but perhaps a bit too old and too regular army to sympathize with *all* that goes on around him. Trapper John, virtually Hawkeye's yes-man, has been replaced by a more mature B. J. Hunnicutt (Mike Farrell), whose responses to the eccentrics, the boredom, and the horror around him have more dramatic potential. Self-righteous, cowardly, lecherous, professionally incompetent Frank Burns has been replaced by a disdainful, brilliantly competent, self-appointed Boston aristocrat, Charles Winchester (David Ogden Stiers). And the balance and interplay of characters have been enhanced.

"Barney Miller," on the other hand, seemed content to let its once intricate casting and baroque plotting deplete gradually without replacement. True, Steve Landesberg was brought in when Gregory Sierra departed, and James Gregory has occasionally come back, but Barbara Barrie was written out, Abe Vigoda is gone, Jack Soo is now gone. The show seems to be cutting down by attrition, rather like the New York Police Department itself in this time of budget crisis.

The new champion of casting checks and balances is surely "Taxi," which is perhaps the effort of producer James L. Brooks and company to show that they could do for a bunch of New York lowbrows what they did for the middlewestern middlebrows that surrounded Mary Tyler Moore. If I caution that they have not entirely succeeded, it might be rather like pointing out churlishly that Fred Astaire did only one *Top Hat*, Leo McCarey only one *Awful Truth*, W. C. Fields only one *It's a Gift*. Anyway, Louie (Danny DeVito), a character of absolutely hilarious and absolutely unredeemable meanness, pettiness, and spite, is memorable enough creation for any show to have.

Coming Back

The "New Dick Van Dyke Show" arrived in the fall of 1971, managed a couple of funny segments, garnered no high ratings, went through a change of format and locale, and died.

Mary Tyler Moore, having been the centerpiece in a memorable comedy of relationships in the 1970s, then tried a variety show which went through a change of format and died.

The set-up, the formula, for Van Dyke's return seemed to be tried and true. He worked in TV (as the host of a talk show). He had a wife, a small son, and this time also a college-age daughter (a chance for another range of crises), and the second bananas, Fannie Flagg, Marty Brill, were again verbal clowns, et cetera.

But something was wrong, and the format could not sustain the resources of its lead comedian. The something that was wrong was, I think, virtually the same thing that was wrong in Miss Moore's variety show—and the same thing that led to the rapid decline of Jerry Lewis.

In the old Van Dyke series, the boss, Alan Brady, overbearing, demanding, capricious, was always there. And he encouraged an element of engaging reticence, and quite human cowardice in Van Dyke's character. Put a comic like Dick Van Dyke in charge or alone in center stage ("in one" as performers have it) and he gets a little pompous—to push the point a bit, try to imagine a take-charge Stan Laurel or a responsible Bert Lahr.

My purpose in raising the question is not to second-guess Van

Dyke's failure but, as I hint above, to raise a question that the pursuit of comedy itself raises often.

The Brady character was, in effect, a bossy straightman (and always a bossy off-screen threat) to Van Dyke's high-level clown. And most clowns need straightmen of one sort or another in order to be clowns. Those who lose them, or who thrust themselves to center stage—into a quasi-authoritative spot—well, often they can't be funny any more, or they can't be funny in the same way.

Thus Jerry Lewis. And thus Mary Tyler Moore in her variety show. When she had to do the chores of an emcee the polite reticence, the sometimes awkward always engaging shyness, the deferential reticence of Laura Petrie or Mary Richards necessarily disappeared, and the leading lady was neither comic nor even particularly likable.

So how did a true clown like Carol Burnett manage to stand "in one" and relate both to the camera and the studio audience in her nightly introductory segments? She waited for the questions, the comments, and the requests that came from the audience, and at every possible opportunity turned them into straight lines against which she clowned unashamedly.

Mary Tyler Moore's sitcom series was indeed a high accomplishment for the tube and a step forward for TV comedy. It is not enough to say that the writers provided fresh, clever, fanciful and sometimes provocative gags. It is also not quite enough to say that the intricate system of comic types and the complex cast of foils and counterfoils the show was very carefully wrought and sustained. They also took chances with it.

Betty White's Sue Ann was not the frustrated spinster who wanted to get married. She wanted to get laid, and if the laying could help her career, so much the better. ("How do you think I got *this* job?" she once demanded of Mary.) They even let Ed Asner's Lou Grant tumble into her bed in an alcoholic moment—and survive the experience. And Cloris Leachman's Phyllis was a married mother who often behaved like an officious, lifetime spinster. Valerie Harper's Rhoda wanted a man too, but was always a bit too competitive to get one, or to keep one. Mary never seemed sure whether she wanted one permanently or not.

They had Ted Knight's Ted Baxter behaving like such an incompetent on camera as to strain credibility—even the credibility we allow buffoons. And the wisecracks of Gavin McLeod's Murray reached such a sustained level of simpering bitchiness that they had to reassure us that he really did have a beloved wife and kids in the suburbs, and have him—momentarily, sentimentally—fall for Mary.

The highest accomplishment of the series was that the producers and writers generally resisted the inevitable sitcom temptation to put any good gag in anyone's mouth for the sake of a quick laugh and another minute of playing time. There were few lines spoken and few attitudes taken that were not in character, did not reveal character—and that sometimes tested and developed character. Not only were the characters allowed to grow, that growth was not imposed on them but integrated with their comic relationships and conflicts, and revealed through the very gag lines that amused us for their own sake.

Well, do they, can they come back? They can and they do of course, and for James Garner on "The Rockford Files" they do just about triumphantly. The series was evidently a slight reworking of the earlier Darren McGavin "Outsider" series, tailored for the highly resourceful Garner. It came about as a mid-season measure to reintroduce Garner on TV. An earlier effort called "Nichols," a TV variant of Garner's 1969 film, *Support Your Local Sheriff,* was cancelled in mid-season.

Supported by lively and interesting secondary characters (a bunch of independent, self-determined people indeed!), and some generally interesting scripts,* Garner's Rockford was an idealistic,

---

* In passing, perhaps the most consistently good suspense scripts ever were those on "Hawaii-Five-O." They used well-conceived variations on the standard mystery, suspense, chase, organized crime, super-villain, interesting impostor, and wayward son or daughter devices, and, it seems to me, the scripts accomplished more than any other element in sustaining that long-running show.

sympathetically lazy, prudently vulnerable private eye, constantly put in the position of knowing that a penchant for self-preservation can be the better part of valor. He relished a good con, and good impersonation, or an effective white lie in the name of justice. Still, Rockford always understood that the key to it all was a shrewd but never cynical acceptance of the avarice and venality of his fellow men.

Rocky's foils were interesting variations on elements of his own character, pushed almost to their limits. His father was reluctant, reticent, conservative—but usually out of well-meaning innocence. Angel, Rockford's onetime cell mate (Rockford had been imprisoned, but was pardoned) was a con man comically conceived without scruples or loyalties, shrewd but without very much intelligence, and certainly without any class. And Rocky's female attorney was idealistic strait-laced, but (always on the lookout for underdogs) exasperatedly devoted to Rockford.*

Garner is an almost perfect light-comic lowbrow—almost perfect for TV, that is: intimate, unforced, entirely unpretentious, knowing where the camera sits and where the audience will sit, knowing that in order to play it as close and casual as one must on TV, the actor must play his character as close to himself, or a part of himself, as he can. James Garner the actor is full of the knowledge of what he can do and can't do, of what he may risk or shouldn't risk, and of how to get almost any effect in his own way. He is as commendably reluctant to do anything he does not understand and does not feel as Robert Mitchum. And his acting shows us, moment after moment, that life is a very serious, even dangerous, and a richly comic experience all the time.

In passing, I'd like to call attention to a wonderful black comedy,

---

* It is interesting that when the Rockford producers tried a follow-up show, "Ten Speed and Brown Shoe," they split up Rockford's character, giving his idealism to an innocent "Brown Shoe" (Jeff Goldblum), and the proclivities and talents of a con man to "Ten Speed" (Ben Vareen)—in effect, they offered a pair of half-heroes, and it didn't work.

late in Rockford's run, about some Eastern mafia hoods, totally unable to function in a California, where life is slower and more innocent, and most people actually seem to *like* their jobs. The hoods invoked such concepts as "loyalty" and "standards" on behalf of the good crime-syndicate life back East.

Index